IDEOLOGY AND CINEMATOGRAPHY IN HOLLYWOOD, 1930–39

Also by Mike Cormack

IDEOLOGY

Ideology and Cinematography in Hollywood, 1930–39

Mike Cormack
Lecturer in Film and Media Studies
University of Stirling

St. Martin's Press New York

Scholarly and Reference Division,
St. Martin's Press, Inc., 175 Fifth Avenue,
New York, N.Y. 10010

First published in the United States of America in 1994

Printed in Hong Kong

ISBN 0–312–10085–X (cl.)
ISBN 0–312–10067–1 (pbk.)

Library of Congress Cataloging-in-Publication Data
Cormack, Michael J.
Ideology and cinematography in Hollywood, 1930–39 / Mike Cormack.
p. cm.
Includes bibliographical references and index.
ISBN 0–312–10085–X (cl.) — ISBN 0–312–10067–1 (pbk.)
1. Motion pictures—Political aspects—United States.
2. Cinematography. I. Title.
PN1995.9.P6C67 1994
791.43'0973'09043—dc20 93–17636
CIP

Contents

Acknowledgements

Thanks are due to John Izod who supervised the doctoral thesis on which this book is based, and to John Caughie and Phillip Drummond, the examiners of the thesis, whose suggestions helped its transformation into a book. Particular thanks are due to Chris. Without her encouragement the writing of both thesis and book would have taken much longer and been much more onerous.

1 Introduction

For the United States, the thirties began on 29 October 1929 when Wall Street crashed and ended at 1 p.m. on 7 December 1941 when Japanese planes struck at the American Pacific Fleet in Pearl Harbor, Hawaii. Between these two traumatic events stretched a decade dominated by the economic depression which followed the crash and which was only exorcised finally by the change to a war economy in the forties. Politically the period was dominated by Franklin Delano Roosevelt, elected as President in 1932 with his promise of a 'New Deal' and who stayed at the White House until his death in 1944. Roosevelt's far-reaching and dramatic changes in American economic and social life make this a remarkable decade in the nation's history.

But this was also the decade of Hollywood, a 'golden age' of mass entertainment, beginning perhaps as early as October 1927 when the success of *The Jazz Singer* heralded the end of the silent era and closing with that quintessential Hollywood film, *Gone With the Wind*, whose intense Technicolor images and heart-rending orchestral score were premiered in December 1939. The contrast between the films of this 'golden age', variously characterised as escapist, simplistic or infantile, and the harshness of the economic climate, not to speak of the upheavals in the social fabric, are striking enough to call for investigation. The relationship between film and society is nowhere more intriguing than in this decade in which Hollywood established the pattern for the sound feature film – a pattern which was to be followed by other film industries in other countries.

Several writers have commented on the link between the content of the films of this period and their social environment, notably Lewis Jacobs in *The Rise of the American Film* (the first edition of which was published in 1939) and Andrew Bergman in his 1971 book *We're in the Money: Depression America and its Films*. What these writers do not account for is the style of the films, and yet, if the socio-economic environment has had an impact on the content, it seems reasonable to assume that it has also had some kind of impact on the style as well. This book will be an investigation of precisely this point – the socioeconomic determinations of cinematographic style. It will be an attempt to answer the question 'Why do Hollywood films of the thirties look the way they do?'

EXPLAINING CINEMATOGRAPHIC STYLE

Traditional explanations of cinematic style tend to fall into one of two general categories – explanations based on individual creativity and explanations based on technological change. At their simplest the former are reduced to the 'great man' theory of history. Historical change comes about because of the actions of great men (the classic examples being Alexander the Great and Napoleon). A rather more respectable version is common in histories of the arts. Stylistic and artistic change is caused by the actions of a few creative individuals who set the pattern for those who follow. This account is perhaps less common now in film histories than it used to be, but can still be seen to be influential. Books on cinematographers often set up a pantheon of great men whose pioneering work is the cause of stylistic change – men such as Billy Bitzer, William Daniels, Lee Garmes, James Wong Howe, Gregg Toland, John Seitz, Leon Shamroy, Haskell Wexler and Gordon Willis in Hollywood, and Raoul Coutard, Vittorio Storaro and Nestor Almendros in Europe.

Charles Higham's 1970 book *Hollywood Cameramen* is one such work, seeing innovation in purely personal terms. 'It was in the mid twenties, at about the time of the development of panchromatic film in 1924, that strongly individual styles began to emerge. John F. Seitz began to use Rembrandt's north light in films made with Rex Ingram, and the results were markedly individual. Victor Milner and Lee Garmes took up the mode, and transformed it into separate and personal styles.'[1] Despite the mention of panchromatic film, the emergence of varying styles is linked only to individual personalities. The early thirties is seen by Higham as a time when individual cameramen of talent struggled to overcome the problems set by the introduction of sound.

> Yet a few cameramen, in some cases aided by their directors, did succeed in overcoming this crippling situation. Von Sternberg and Garmes created the rippling shadows and shining burnouses of *Morocco*, while Garmes's north light shone beautifully on the face of Dietrich. At Warners, Ernest Haller helped to establish a traditional low-key look in films of the calibre of Dieterle's *Scarlet Dawn*. Miraculously, cameras fought to follow a drunken Fredric March in *The Royal Family of Broadway*, travelling from a train through Penn Station to the station ceiling to take in a sudden flight of doves in Mamoulian's marvellous *Applause*.[2]

This view of cameramen struggling against the technology to express themselves is analogous to Andrew Sarris's view of directors struggling against the studio system.[3] Personality is proved by the overcoming of obstacles.

Leonard Maltin continues this approach in *The Art of the Cinematographer* (first published one year after Higham's book). 'A handful of creative artists continued to shine during the transitional period of the talkies, but not until the early 1930s did the cameraman regain his former stature, and begin to take up where he had left off in 1929.'[4] This was the period when 'imagination and creativity returned to filmmaking, and a new crop of outstanding cameramen established themselves.'[5] Maltin is more sensitive to other factors than Higham (noting, for example, the importance of Gregg Toland's contract with Samuel Goldwyn in allowing him a certain amount of freedom)[6] but still sees everything ultimately in individual terms. Even studio style is reduced to one cameraman's work – 'You knew it was a Warner Brothers movie by . . . the photography of Sol Polito.'[7] The fact that Higham attributes Warner Brothers style mainly to Ernest Haller merely emphasises the problems of this approach (and neither of these authors even mentions Dev Jennings whose striking low-key cinematography on *The Public Enemy* is described below in Chapter 4).

Understandably perhaps, this approach is strongest in books devoted to one individual. Todd Rainsberger's study of James Wong Howe is the major example. 'Inasmuch as the cinematographer usually controls the visuals, his own method of filming takes on an importance far beyond mechanical reproduction. The individual way in which he manipulates his raw materials expresses his own concept of the script – a commitment of his ideas to film – in the same manner that an artist uses paint to portray ideas on canvas.'[8] Not all cinematographers, however, achieved this level of expression – and certainly not in the studio years. 'But only those cameramen who could impose their vision through sheer force of will achieved any sense of individual style. One such cinematographer was James Wong Howe.'[9] Howe's use of low-key lighting and a kind of realism based on an attempt at source lighting (in which lights on the set were placed so as to reproduce natural lighting effects) mark him, according to Rainsberger, as one of these visionary leaders of style, setting the standards which lesser cinematographers would follow.

The problems with this view are fairly obvious. It does not explain why any specific innovation became widespread. The fact that an individual innovates in a particular way does not reveal why some innovations are taken up and become part of the dominant style, while others are not. In addition to this, the individualist argument does not actually explain anything. It suggests that stylistic change is due to individual creativity but this serves only to block the question – it does not explain why a specific individual innovates at a specific time. To put it down to the mystery of 'creativity' is simply to avoid dealing with the question. This is not to argue that individual cinematographers were not important. Garmes, Howe and Toland, for example, were clearly working at a different level from their more routine studio colleagues, but to isolate them from any environment and to suggest that no further explanation can be given beyond their 'genius' is simply mystification.

An alternative group of arguments attempts to explain stylistic change by technological progress. A particular change in technology (such as the introduction of sound) will lead to a particular change in style (such as a greater use of reframing). The changes in technology are simply the result of the inevitable progress of scientific knowledge. As long as 1974 Raymond Williams demolished this argument: 'Technological determinism is an untenable notion because it substitutes for real social, political and economic intention, either the random autonomy of invention or an abstract human essence.'[10] In other words it is a theory which ignores the fact that every person and every invention is situated in a specific social, political and economic context.

A slightly more sophisticated version combines technological progress with a natural evolution towards greater realism. Particular technological changes are fuelled by the urge to realism and when changes occur, they lead naturally to particular uses. The developments in filmstock during the thirties (noted below in Chapter 6) were, according to this argument, the result of the studios' desire for more realistic films, and when the new faster stocks became available they resulted in other changes in the same direction, such as deep focus. Patrick Ogle's account of the origins of deep focus is a classic version of this argument. Ogle describes the various technological changes which took place in the twenties and thirties, including developments in film stock, lighting and lenses, some of which had unintended consequences. 'With the at least temporary discontinuance of the use of arc lamps owing to noise problems on the newly built sound stages, American film-making entered a period of heavily diffused images, soft tonality, and shallow depth of field that was to characterise Hollywood films until into the later 1930's.'[11] Thus the introduction of sound (to add realism) results in a particular type of cinematography. This is changed only when other changes (silent arc lamps, faster film stocks, coated lenses) occur. These changes again lead to a more realistic style.

This argument has, of course, a famous progenitor in André Bazin who argued that the technical problems of the cinema were essentially solved by 1930 and that subsequent changes were guided by what he called 'the myth of total cinema'.[12] This myth was that cinema should achieve 'an integral realism, a recreation of the world in its own image, an image unburdened by the freedom of interpretation of the artist or the irreversibility of time'.[13] For Bazin, the deep-focus, long-take style of Orson Welles and William Wyler best achieved this aim. Bazin puts the emphasis on the desire for realism, whereas Ogle puts it on the technological changes themselves, but both see an interplay between the two factors.

Raymond Williams' refutation of technological determinism serves as effectively against Bazin and Ogle as it does against Higham, Maltin and Rainsberger. It is worth adding to this that even the arguments of Bazin and Ogle cannot explain why particular techniques become widespread. A mobile subjective camera would appear to be a useful contribution towards realism yet it is a device which has never become popular for anything more than the occasional special effect. Colour would seem to be an obvious step in the direction of realism and yet for a long time its major use was in the most unrealistic genres, such as musicals and westerns. The point is, of course, that the conventions of realism are variable. The argument that film technique is continually aiming towards greater realism is only convincing because the concept of realism is so empty of content that virtually any technique can be linked to it. The need to provide the audience with something new, while at the same time attempting to limit the possibilities of meaning to an ideologically safe zone, is a more convincing explanation of successful cinematic innovation, as will be demonstrated in the following chapters.

It is worth adding that Bazin and Ogle are both extremely selective in their accounts of Hollywood film style. This is obvious in Bazin's comments on thirties films but is also present in Ogle's essay. As will be be clear from the above quotation, he sees no significant stylistic change between the introduction of sound and the development of deep focus, all films in between being described as having 'heavily diffused images, soft tonality, and shallow depth of field'. The various styles of such films as *The Public Enemy*, *Viva Villa!*, *Bride of Frankenstein* and even *Gone With the Wind* are unrecognisable from this description.

More recent attempts at describing and explaining the style of thirties Hollywood films have not been notably more successful than those of the earlier writers. Barry Salt, for example, has deliberately avoided any kind of explanation, preferring simply to list stylistic details. A more sophisticated approach can be found in Bordwell, Staiger and Thompson's *The Classical Hollywood Cinema*, although even here there is a tendency towards mere description. The authors limit explanation to a combination of one or more of three factors: (1) production efficiency, (2) product differentiation, and (3) adherence to standards of quality. The third of these categories causes particular problems. As examples of such standards the authors cite 'progress toward better storytelling, greater realism, and enhanced spectacle'.[14] When these vague, evaluative phrases are unpacked it becomes clear that they are seen in a near-Bazinian sense of an evolutionary movement towards an ideal of cinematic expression. The meaning of 'better storytelling', 'greater realism' and 'enhanced spectacle' can only be specified by making assumptions about human nature, about society and about the role of cinema. Once this is seen, it becomes clear that ideology is lurking behind these criteria and that a theory of ideology is needed to overcome this teleological narrative of film history.

But the problems of Bordwell and his co-authors do not end there. They argue that there is a constancy of style in Hollywood films throughout the period 1917–60 and they describe this consistency as 'the classical stylistic paradigm'.[15] Differences within this period are mainly differences in the uses of specific technical devices. Underlying these differences is a continuity in what they term the 'stylistic systems' of narrative logic, cinematic time, and cinematic space.[16] The continuity of the classical style rests on these systems and on the unchanging relationships between them. The trouble with this approach is that it consistently underestimates and undervalues the stylistic variety of Hollywood films. By reducing style to underlying abstract systems of narrative, time and space, attention is deflected away from the surface of the text and so the significance of change at this level is missed.

The authors of *The Classical Hollywood Cinema* are not the only ones to make this mistake. A very different writer, Noël Burch, has argued in favour of the notion of an unchanging 'institutional mode of representation' in Hollywood since the introduction of sound.[17] This mode consists of narrative elements (the causal chain of events) and diegetic elements (the simulation of spatial and temporal coherence). As with the model of Bordwell, Staiger and Thompson, Burch's generalisations flatten out differences in style, assimilating them all to one pattern. The advantage which Burch's approach has over that of Bordwell, Staiger and

Thompson is that in it ideology does have a clear role to play. As he puts it, 'the pressures – economic, ideological, and cultural – that were eventually to create, first in the United States, then throughout the Western world, the conditions for the triumph of the Institutional Mode exerted themselves on the cinema as soon as it was born'.[18] Unfortunately this broad sweep which sees all popular film as the result of a single bourgeois ideology (and Burch is certainly not the only writer on ideology and film to take such a broad view) loses the detail of ideological variation between different countries and different historical periods.

The argument in the chapters to follow will be that there were significant stylistic changes in Hollywood films during the thirties and that these changes can only be fully explained by considering the specific ideological forces at work on the films. Undoubtedly there were many strong similarities in films throughout the decade, but this should not deflect attention from their very real variations.

THE STUDIO SYSTEM

The 'flattening out' of film history of this period is also apparent in descriptions of the mode of production in Hollywood in the thirties. Film production at this time was dominated by a small group of vertically integrated companies, that is, companies which produced films, distributed them, and also owned many of the biggest and most important cinemas in which they were shown. Thus they had complete control over their market. These companies (the 'majors') were MGM, Warner Brothers, Paramount, Twentieth Century-Fox and RKO. Behind them (and cooperating with them to control the market) were the three 'minors', two of which (Universal and Columbia) were producers and distributors but did not own cinemas, and one of which (United Artists) was only a distributor, dealing with the top rank of the independent producers (including such figures as Samuel Goldwyn, David O. Selznick, and Walt Disney). Beyond these eight market leaders were a large number of small producers of cheaply-made films ('B-movies') who depended on the major distributors and exhibitors to get their product into the cinemas. The best-known of these so-called 'Poverty Row' companies included Republic, Monogram and Mascot.

This hierarchy of studios was enforced by a set of practices which would be declared illegal in the late forties. Blind booking (exhibitors having to book films unseen), block booking (exhibitors having to book films in batches, thus taking bad films along with the popular star vehicles) and zoning (cinemas being organised into a hierarchy, those at the top getting new films more quickly than those further down the list and thus being protected from neighbourhood competition) were some of the means by which this economic oligopoly survived.

But to give the outline of the economic system in this way should not obscure its instability. During the thirties some majors became bankrupt, most had important changes of top personnel, and most made significant changes in the kind of film they made. All these changes tend to be downplayed in the standard histories. This can lead to the assertion, for example, that Universal was best-known for horror films throughout the thirties, despite the fact that from 1936 to 1939 its best-known films were Deanna Durbin musicals. Part of the instability was the continual pressure against the system by talented individuals who wanted more freedom in their work. By the forties this pressure had resulted in stars such as James Cagney and directors such as John Ford setting up their own production companies. The studio system needed their talents, but inevitably frustrated some of their ambitions. The system was also made insecure by a more obvious pressure. The studios were attempting to provide films which were new but which used patterns of proven success. A factory system was being used to create a product which was totally unlike any mass-produced factory product. This was bound to produce problems as the public's desire for novelty collided with the studio's desire for consistent financial profit. The system was given the facade of stability only by those distribution practices which would be declared illegal in the forties.

The cinematic history of the thirties, then, displays an intriguing mixture – social and economic upheaval, a flourishing film industry which was inherently unstable and a continually changing range of stylistic practices. The aim of this book is to trace the interactions between these factors.

DEFINING 'STYLE'

At this point it is important to be more specific about exactly what is meant by cinematographic style. It cannot easily be abstracted from other filmic elements such as *mise-en-scène*, performance or script. The attempt to study it alone is not to be taken as an assertion of its autonomy; it is merely an analytic convenience. The main elements to be considered will be the four variables coming under the control of the director of cinematography which varied most significantly in films of the thirties – camera movement, camera angle, camera distance and lighting. These are taken to be the principal (although not the only) components of cinematographic style. Style is simply the way in which these elements are used in a film. Any shot must have some kind of lighting and some placing of the camera. Thus cinematographic style enters into every shot. The style of a film is the system which describes the overall use of these elements. In this usage style is a descriptive term, not an evaluative one – it is not something which only 'good', 'stylised' or 'art' films have.

METHOD

The chapters which follow develop a theory to link ideology with visual style. Chapter 2 will give an account of ideology and the implications of this account for the study of film, and will then develop a method for linking cinematographic style with its ideological determinants. This method will be based on the two filmic systems of signification (how visual style signifies in an ideological framework) and subjectivity (how visual style attempts to construct viewing subjects in ideology).

The remaining chapters will work through the thirties in Hollywood by juxtaposing detailed analyses of four films with more general discussions of the early, middle and later thirties. The four films are not chosen to give a complete account of the stylistic changes of the decade but rather to illustrate certain key elements. In Chapter 3 *All Quiet on the Western Front* (Universal, d. Lewis Milestone, 1930) will illustrate Hollywood style at the point of change between the late silent and the early sound film. In Chapter 5 *Dr Jekyll and Mr Hyde* (Paramount, d. Rouben Mamoulian, 1932) will illustrate the furthest reaches of mainstream experimentation in the early thirties. *The Awful Truth* (Columbia, d. Leo McCarey, 1937) in Chapter 7 is chosen to represent the more restrained style of the middle of the decade and in Chapter 9 *Dead End* (Goldwyn, d. William Wyler, 1937) will illustrate the changes which were beginning to occur towards the end of the period. The intervening chapters will generalise the argument by considering a wider range of films throughout the thirties and by relating the changes apparent in these films to the developing socioeconomic context. The aim is to avoid the two common pitfalls of historical analysis – generalisation from too small a sample and superficial mention of too many films. Chapter-length analyses of four films, shorter but still substantial analyses of another eleven films in Chapters 4, 6, 8 and 10, and beyond that briefer mention of a number of other films, should, taken together, provide a representative and illuminating sample.

2 Ideology and Cinematographic Style

The concept of ideology has had a long history in which widely differing accounts have been defended and in which no single theory has been able to claim outright victory. The contemporary significance of this debate is that it makes it important for any writer using this concept to give it a clear definition, but one which is complex enough to encompass the problems raised by recent critiques. The simple definitions which earlier studies on ideology and film used (such as in Comolli and Narboni's famous essay 'Cinema/Ideology/Criticism', in Peter Biskind's book *Seeing Is Believing*, or Bill Nichols' *Ideology and the Image*) can no longer be considered as adequate. Adequacy can only be achieved by articulating together a series of partial definitions which together can cope with the inherent conceptual problems.[1]

THE ARTICULATED DEFINITION

Any stable society will be organised around a preferred self-image. By 'self-image' is meant the way in which the society is described by its dominant groups. The function of this representation is to reproduce its own conditions of existence, in other words to protect the status quo. Thus the social self-image is rooted in socioeconomic conditions. A society undergoing crisis, particularly a crisis over which groups should be dominant, will manifest competing self-images, but one must eventually become accepted as the dominant image or else the society will lose its coherence. These self-images will be formed within a framework of interlocking concepts (around such issues as democracy, liberty, morality, justice, race, gender, religion, nationality, etc.), that is, a discursive organisation. These discourses will structure the linguistic and social practices which provide the means by which any member of that society can produce meaning and thereby communicate with other members of the society. Language is spoken in a particular social context. The interaction between the symbolic system (spoken language) and the social system (the framework in which speech takes place) produces meaning. The conceptual framework and its related set of values mark the limits of acceptable debate. Language users can use language in new ways but if they move far outside the ideological framework they are likely to lose their audience.

The relationship between this framework and the specific mental contents (such as ideas, representations, beliefs, opinions, values, and judgements) which are structured by it should be apparent. Also important is the notion of subjectivity – the way in which ideology attempts to structure individuals as ideological subjects. Each person's self-image is not derived from a value-free, objective essence of selfhood, a centre of will and intentional action. Rather, the self-image is socially-derived (language is, after all, a social phenomenon) and is dependent on heavily-loaded ideological concepts. Individuals can only define themselves in social terms.

Ideology, then, is a process which links socioeconomic reality to individual consciousness. It establishes a conceptual framework, which results in specific uses of mental concepts, and gives rise to individuals' ideas of themselves. In other words, the structures of human thinking about the social world, about themselves and about their role within that world, are related by ideology ultimately to socioeconomic conditions. In addition to this it is worth noting a point made by some theorists who insist on the material existence of ideology in social institutions, which is that ideology exists not just as free-floating ideas and values, but also as institutionalised in the practices of society, such as the legal system, the family and the media.

THE MINIMAL CONCEPT OF THE DOMINANT IDEOLOGY

Although the central stream of thinking about ideology has been within the Marxist tradition, recently some of the central ideas of that tradition have been subject to powerful critiques. Under particular attack has been the notion of the dominant ideology. Some writers have argued that even if there has been a dominant class with a dominant ideology in the past, there is no longer anything which can be labelled as such in developed countries such as those of Western Europe and North America. As James Collins has put it, the notion of a dominant class 'may well be a fascinating methodological fiction, but it only obstructs our understanding of the complexity of the conflictive power relations that constitute our cultures'.[2] Thus any notion of the dominant ideology must, he argues, be rejected out of hand. Other writers have questioned whether there has been such a thing in the past – or at least whether it has had any importance. Abercrombie, Hill and Turner, in their studies of feudal, early modern and contemporary societies, find the idea lacking the significance which earlier writers had assumed, and in their most recent update of these arguments have not only added a broader range of examples but have added the idea that the texts of popular culture (such as films) have a detectable ideological effect is indefensible.[3]

In this situation, it is worthwhile making two points. The first is that the assertion that a dominant ideology exists and is significant does not imply that everybody thinks the same way, or even that everybody in authority thinks the same way. All that is necessary is sufficient agreement concerning the structure of the socioeconomic system. This agreement may even be simply at the basic level of acceptance, rather than anything more positive. Abercrombie *et al.* are concerned with a strong notion of the dominant ideology but the weak or minimal notion suggested here escapes the main force of their criticisms. Since the function of ideology is to reproduce the conditions of its existence, such a minimal level of agreement is not only all that is necessary, but may actually be more effective than any stronger form of agreement. As capitalist countries have gradually discovered over the last century, allowing a certain amount of disagreement over political and social aims is conducive to the acceptance and continuation of the overall system.

One of the most dramatic examples of this is found in the period under study. As the Depression hit America in the early thirties, there was a great deal of social unrest and some at least feared that society was under threat, particularly as President Hoover refused to intervene in the economy. When Roosevelt was elected on a landslide majority in 1932 he seemed to promise a revolution in social and economic policy. Yet, with hindsight, it is clear that this 'revolution' in fact stabilised the system by making some fairly minor (although important) changes, such as introducing public works spending and social security. The socioeconomic structure of the country was unchanged while at the ideological level some expansion had taken place. By the middle and later thirties, the dominant beliefs and values of America were clearly in evidence and unthreatened, whereas in the early thirties the economic crisis had led to an ideological crisis.

The second point to make about the dominant ideology is that the consequences of there being no dominant ideology would be not just social upheaval, but social disintegration. The concept of a dominant ideology is simply that of a consensus of beliefs which are held to be commonsense and which function as the principles of social cohesion. A society without such a dominant set of beliefs and practices is a society in crisis. In fact, of course, the usual situation is for a certain amount of contestation to take place, both within the broad church of the dominant ideology as different groups come into conflict (different genders, different races, different regions, even different professional groups), and outside it as oppositional groups attempt to influence the boundaries of the dominant discourses. It will not then be surprising to find that cultural products will frequently be the site of ideological conflicts, even (perhaps especially) those products which are aimed at a mass audience.

The dominant ideology, then, is a necessary part of any stable society, but is not a monolithic, necessarily coherent force. Rather it is a mass of different forces, different voices, which are aligned and realigned as circumstances change (and here the theory of ideology gets close to pluralist accounts of society, although maintaining the notions of dominance and socioeconomic determination). As Terry Eagleton has put it, 'Ideologies are usually internally complex, differentiated formations, with conflicts between their various elements which need to be continually renegotiated and resolved.'[4] All that *is* minimally needed is for these alignments and conflicts to take place within one framework (which for many countries might be defined by the concepts of parliamentary democracy, government sovereignty, patriarchy and a capitalist system based on property rights and unending economic growth) and for oppositional groups to be marginalised in every way possible.

HEGEMONY

This notion of the dominant ideology links up with the concept of hegemony, as developed by the Italian philosopher Antonio Gramsci who argued that in order to retain power, the dominant group in society will attempt to incorporate some of the aims of subordinate groups, thus giving these latter groups a reason for supporting the dominant group. Leadership is retained by allowing, and indeed promoting, the expression of some of these aims. Thus the hegemony of the dominant group is maintained not by force but by consensus. The importance of this concept is not just that it seems to describe rather better than traditional Marxist ideas how modern liberal-democratic societies function, but also that it implies that ideological struggle and contestation is always taking place, as the different groups in society articulate their endeavours in ever-changing ways. This in turn suggests that cultural products aimed at a mass audience, such as Hollywood films, will in some way reflect this continual struggle. Rather than looking for simple, straightforward ideological messages in films the analyst should be aware of the likelihood of finding evidence of conflicting interests, contradictory discourses and unresolved struggles. To take one well-known example, *Gone With the Wind* has appealed over a period of many years to people of all social classes, both genders, and many nationalities. It would therefore be extremely surprising if it turned out to have a simple, unambiguous ideological structure.

It is worth noting another advantage of the Gramscian account. It allows for society to be seen as more than just competing social classes. Conflicts based on gender, race, sexual orientation, age and region can be made to fit into a theory of ideology if hegemony is considered. Each of these aspects can be seen as inflecting class structures and creating a large number of potential divisions. Since these are all features which play an important part in popular films, this is obviously an important advantage. Some writers have taken this point further and argued that ideology needs to be detached from any necessary relationship to class in favour of a more generalised relationship to structures of power of any kind. Thus John B. Thompson has defined the study of ideology as the study of 'the ways in which meaning serves to establish and sustain relations of domination'.[5] He makes clear that such relations are very varied. 'We live in a world today in which class domination and subordination continue to play an important role, but in which other forms of conflict are prevalent and, in some contexts, of equal or even greater significance.'[6] Two points can be made in answer to this. The first is that it is within a structure of social and economic stratification that other forms of domination take place. The basic necessity of economic survival means that the fundamental structuring of society is economic (and even many non-Marxists would accept this), and so the social forms based on that deserve to be regarded as the primary structures of society, with gender, race and other structures as secondary in that they are dependent on the former. The second point to be made is that even if Thompson's account is accepted as accurate for contemporary societies, it is less convincing when more traditional industrial societies (such as the USA in the thirties) are considered. Having said this, it is clear that issues of race and gender in particular have an important relevance when considering Hollywood in the thirties.

DETERMINATION

Much debate about ideology has turned on the question of determination. The traditional view sees ideology as directly determined by economic realities. The *superstructure* of beliefs and values rests on the *base* of the economy. To avoid such simplicities it can be noted at once that these concepts of superstructure and base are abstracted from the social world. They are a metaphorical representation of a complex reality. If they are seen rather as analogous to the vertical and horizontal frames of the social formation, then their interdependence will become clearer. Both are necessary for a society to exist, but there is no obvious reason why causation should only run from base to superstructure, rather than from superstructure to base or indeed across the superstructure. Even if, as has been argued, the economic level is the fundamental structuring level of society, that does not mean that every aspect or change of the superstructure of beliefs need be derived immediately from economic factors.

This suggests that ideology has what has been called a *relative autonomy* from the economic base. It is not completely autonomous (after all ideology does ultimately owe its existence to economic differences) but neither is it necessarily completely determined. Relative autonomy, however, is not a single value. Rather it covers a continuum between two extremes and how much autonomy existed at any particular period remains a matter for investigation. Even commercial film studios in Hollywood cannot be said always to have the same degree of autonomy. The varying historical conditions in which such studios found themselves must be considered as well. The question of how much autonomy any institution has had at any point in its life will be discovered only by studying ownership, control, content and the individual responses of the people involved in the institution.

POPULAR FILM

The potential complexity of the problem of analysing popular film should now be evident. The economic aim of popular film is to make money. It does this by gratifying the audience's self-image. Since that audience is a complex group and since self-images can be constructed in many different ways, it follows that the ways in which ideology will gain expression in film are likely to be many and varied.[7] Clearly such basic parts of the content as dialogue, action and character will have important roles. After all, ideology directly concerns what people say and do, and what kinds of character are seen as good and bad. But ideology can affect films in more complex ways. Narrative structure is important, in particular how a story ends, and how loose ends are tied up. An open ending can leave more than just the story undecided. If a story can be seen as the working out of some problem, then the end will be particularly important. Intertextuality (that is, how a film is related to other texts, including other films) is also important. If a film is seen as a generic film, such as a western or a musical, then this may affect how it is understood. Similarly if a film is seen primarily as the work of one individual, whether a director such as John Ford, a producer such as David O. Selznick, or a star such as Mae West, then this also may affect how it is received, bringing some aspects into prominence and relegating others to obscurity. Also important to consider in films are the more technical elements whose relationship to content is not so obvious. Set decor, music, special effects, editing and, of course, cinematography can all be presumed to have some ideological role to play since they all mediate in some way the cinematic experience. This book will be concerned primarily with the last of these technical elements – cinematography.

THE AUDIENCE

A main reason for studying the popular media is because they are presumed to have some kind of effect on the audience. However, the study of media effects on the audience is notoriously difficult to carry out with any certainty. Despite the claims of populist politicians and moral compaigners, evidence of direct effects on contemporary audiences has not been proved with any academic rigour. Even the notion of what an effect *is* can be disputed. Immediate short-term behavioural effects, longer-term attitudinal effects and unconscious changes are all very different and must be studied in appropriately different ways. This makes the study of the effects of popular film in the thirties all the more difficult. Accordingly this book is not to be taken as making assertions about audience reaction. It is rather to be seen as a study of a large group of films with the intention of working back to see how they were affected by ideological factors. How these films in turn affected their viewers is a very different study.

This is particularly important since many audience researchers now emphasise the room for manoeuvre that audiences have and the ability of the audience to read films in ways very different from that intended by the makers. And of course it is important to remember that the 'mass' audience was never a unity. Even in the USA in the thirties, although the film-makers frequently claimed to be making films for the whole audience, the actual practices of advertising, distribution, and generic film-making make it clear that different films were addressed to different segments of the audience. Some films were aimed primarily at women, some films played mainly in the cities rather than in the rural cinemas, and some films did not reach particular regions such as the South. The mass audience should be seen as an amalgam of many different, smaller audiences (and even these smaller audiences were loose agglomerations).

Having said this, some assumptions about the audience have to be made in any discussion of textual meaning. In *S/Z* Roland Barthes discusses the codes of meaning which interweave in any text. These codes are not strictly bounded paradigmatic sets, but are rather, 'a perspective of quotations, a mirage of structures . . . so many fragments of something that has always been *already* read, seen, done, experienced'.[8] In other words, the codes are aggregations of meaning, histories of contexts and interpretations. John B. Thompson has put this another way by describing media products as 'pre-interpreted domains', that is, before the analyst looks to the texts they have always already been interpreted by audiences and in fact the texts themselves are interpretations of the social world.[9] Such an approach allows us to discuss meaning without being tied down either to denotation alone or to specific audience reactions. Barthes argues that there are five such codes at work in texts: the hermeneutic (the code of the enigmas and resolutions of the text), the semantic (the code of the attribution of characteristics and qualities), the proairetic (the code of actions and causal sequences), the symbolic (the code of connotative and mythic references) and the cultural (the code of references to bodies of accepted knowledge). Although all five have relevance in ideological analysis, the last two are of particular importance when dealing with the analysis of the visual language of cinematography. Kaja Silverman notes how important the symbolic code is for the dominant cultural order and how Barthes describes its working in terms of oppositions: 'Indeed, it could be said that the symbolic code is entrusted with that order's dominant binary oppositions.'[10] The cultural codes also link the specific text under analysis to the larger body of ideological meaning. Barthes describes the statements which use this code as being 'implicit proverbs'[11] which engage a text with a multiplicity of related cultural meanings. To see how these codes work in a text, then, is to see how the text meshes with previously accepted ideological structures.

FROM IDEOLOGY TO VISUAL STYLE

The aim in this study, then, is to examine cinematographic style and how it relates to ideological context, specifically that of the United States in the thirties. If there is any single point from the preceding paragraphs which must be borne in mind in this analysis, it is the complexity of ideology. The dominant ideology may be a complex of contradictory and conflictive voices. Around these may be heard the contesting voices of non-dominant groups. The cultural texts which will be under examination may well show the signs of such contestation due to their producers' eagerness to maximise the audience by appealing to as many different groups as possible. Visual style is particularly important since its less explicit nature makes it ideal both for dominant and for oppositional voices to use. There is, however, a problem here. A method is needed which will allow a link to be made between the context in which a film is made and the visual appearance of that film. Visual style is by its nature less precise in meaning than verbal communication. This is partly because the codes of such elements as lighting, camera movement, angle and distance are analogical, rather than digital, and partly because connotative meaning is a major feature of such stylistic elements and is more difficult to pin down than denotative meaning. Not only does this make the tracing of its origins more difficult, but it means that the understanding of the film by the audience may vary greatly from that intended by its makers. The decoding of the text cannot be determined by its encoding. In addition to this, even the seemingly simple task of describing camera style has its own problems.

THE ELEMENTS OF CINEMATOGRAPHIC STYLE

The four main elements of cinematography to be studied here are lighting, camera angle, camera distance and camera movement – the principal cinematographic variables in Hollywood films of the thirties. Three other possible elements which will not be dealt with in detail are screen ratio, depth of field and use of colour. Screen ratio did not vary at all during the decade. Depth of field did vary but only in a fairly limited way. Broadly speaking it varied with distance, from a shallower field of focus in close shots to a deeper, but not extreme, field of focus in more distant shots. Even by the end of 1939, the use of a deep field of focus in close or medium shots (or, to be more accurate, shots which included a character in close or medium distance as well as another at a further distance) was very rare. Although well-publicised, colour did not play a major role in films of the thirties. The standard history of the Technicolor Corporation lists only 24 Hollywood feature films made completely in the two-colour process and only 29 in the three-colour process in the period 1930–39 (compared with approximately 500 films released each year).[12]

The first step is to elaborate the codes governing lighting, angle, distance, and movement. Christian Metz has defined semiotic codes as 'systems of differentiations' and 'sets of possibilities'. 'If a code is *a* code, it is because it provides a unified field of commutations, i.e., a (reconstructed) 'domain' within which the transformations of the signifier correspond to variations in the signified, and within which a certain number of elements have meaning only in relation to each other.'[13] Metz describes codes as either general cinematic codes which 'are (actually or potentially) common for all films' or particular cinematic codes (also called sub-codes) which 'appear only in certain types of film'.[14] Codes are both paradigmatic (listing mutually exclusive possibilities) and syntagmatic (listing principles of construction).

Metz uses the notion of the 'filmic system' to describe the functioning of codes within a specific text. 'Each film has its own structure, which is an organized whole, a fabric in which everything fits together; in short, a system. But this system is valid only for one film.'[15] This system is an abstraction from the text. 'The system has no physical existence; it is nothing more than a logic, a principle of coherence. It is the intelligibility of the text, that which must be presupposed if the text is to be comprehensible.'[16] He also describes it as 'any underlying organization which is logical and symbolic'.[17] The link between codes and systems is that the latter are actualisations of the former. Many codes will come together in a single filmic system.

Metz is mainly concerned with the system of narrative, but his comments are equally applicable to the systems of cinematographic style. The task in studying the style of the thirties can now be restated as the task of describing the functioning of the sub-codes governing lighting, camera angle, camera distance and camera movement, and the ideological significance of these sub-codes. In addition it will be necessary to study how these sub-codes are used in particular filmic systems with specific ideological significance.

A necessary beginning is the more precise definition of each of the four categories of cinematography mentioned above. Lighting can be subdivided into four categories (following the suggestion of Sharon Russell in her book *Semiotics and Lighting*) – intensity (whether the light is hard or soft), contrast ratio (whether it is high-key or low-key), direction (side light, back light, top light, etc.) and location (shadow in relation to objects).[18] Of these categories, the one which is used most often but which needs further definition is contrast ratio. The terms low-key and high-key are notoriously vague, especially the latter. They also suggest a middle term. Barry Salt offers the following definitions:

Low-key: 'Form of lighting of a film shot which produces an image which is mostly dark.'
Mid-key: 'Form of lighting of a film shot which produces an image which has approximately equal total areas of light and dark distributed over it.'
High-key: 'Form of lighting of a film shot which produces an image which is made up of mostly light tones.'[19]

This allows some precision in the use of these terms, even though their meaning is (necessarily) still fairly loose.

Camera angle can be divided into three categories. The first is the horizontal angle, that is, the angle of the camera on a horizontal plane, around a vertical axis running through the camera at a right angle to the direction of the lens. The second is the vertical angle, that is, the angle of the camera on a vertical plane, around a horizontal axis running through the camera at a right angle to the direction of the lens. The third is the upright angle, that is, the angle of the camera on a vertical plane, around a horizontal axis running through the camera in the same direction as the lens. This gives what is usually known as a 'canted' camera angle. Logically this leads to a fourth variation of angle – the angle of the camera on a horizontal plane, around a vertical axis running through the camera in the direction of the lens. This would lead to considering variations in overhead and underfoot shots. Since such shots are rare in the thirties, they will simply be regarded as variations of the vertical angle.

Camera distance is more straightforward with one range of values from extreme close-up to extreme long shot. There are two complicating factors, however. The first is how to describe a shot in which no human figures appear, that is, a shot in which the normal cinematic measure of distance is removed. This is most obvious when a small object fills the screen. Is a shot in which a four-inch key fills the screen a long shot (since all the object is seen and the shot is therefore comparable to a shot in which all of a person is seen) or a close-up (since the camera is very close to the subject)? The latter choice will be made here, with the recognition of its rather arbitrary grounding. Since very few shots in Hollywood films of the thirties do not contain a human figure, this is a fairly minor point. The second complicating factor is that a shot could contain two people at different distances. Even without the ECU + MS + LS compositions of Orson Welles, shots containing, for example, a character at medium shot and a character at long shot, are not unusual in the thirties. Such shots will here be regarded as containing two distances, rather than any attempt being made to decide which character is the more important one.

Camera movement can be subdivided into four categories. The first is the pan, that is, movement on a horizontal plane around a fixed vertical axis running at a right angle to the direction of the camera lens. The second is the tilt, that is, movement on a vertical plane around a fixed horizontal axis, running at a right angle to the direction of the lens. The third is the track shot, that is, movement of the whole camera on a horizontal plane. The fourth is the elevated shot, that is, movement of the whole camera on a vertical plane. Many shots do, of course, combine these elements of movement. A simple example would be a pan in the middle of a track shot. More complex examples would be shots making use of a crab dolly, a crane or a handheld camera. Such highly mobile shots could be analysed as a series of combinations of the four basic categories. In addition to the type of movement, the speed of the movement is another important variable, as is the extent of the movement. One sub-category which will be referred to is the reframing movement which can be defined as a strongly motivated short pan or tilt. A maximum degree of movement might be suggested to be about 30°. Any larger movement becomes a full pan or tilt (although here again a certain flexibility must be retained).

Having noted the relevant categories, the next step is to describe the opposite ends of the scale of variation in each one, remembering that these are all analogical codes.

Lighting in general: Light/dark (that is, extreme over-exposure to extreme under-exposure).
Intensity of light: Soft/hard.
Direction of light: Diffuse/precise (direction also varies in the straightforward sense of location of the source).
Location of shadow: Absent/present.
Contrast ratio: High contrast/low contrast.
Horizontal angle: Directly in front/directly behind.
Vertical angle: Overhead/underfoot.
Upright angle: Upright/upside-down.
Distance: Very close/very distant.
Pan:
Tilt:
Track:
Elevation:
} All movements vary from stasis to a movement which is too fast to allow identification of the content of the shot.

CINEMATOGRAPHY AS A SYSTEM OF SIGNIFICATION

So far mere description has been used. Next comes the difficult step of deciding what these various elements signify. Each category is based on paradigmatic choice and this makes it possible to work out the signification involved. The well-known commutation test (in which change of one element tests whether the overall meaning changes) will confirm that each variable adds some meaning to the shot. In isolation, however, this meaning can only be described in very vague terms (after all, it is only in combination that full meaning can arise). What can be done is to list what each category contributes to the shot. A more precise meaning (or rather, a more limited range of meaning) will only arise when the element is further defined by combination (not just combination with other cinematographic elements in the shot but also combination with the pro-filmic content and with the rest of the film).

Lighting: Clarity/obscurity (the various elements of lighting are all variations of different kinds within this code).
Vertical angle: Inferiority/superiority (these terms are to be taken in a literal sense in which 'inferiority' simply means 'being below' – the precise connotations of this (physical, social or moral) will be defined by the other elements in the shot).
Horizontal angle: Directness/indirectness.
Upright angle: Balance/imbalance.
Distance: Importance/unimportance (this is simply saying that whatever fills the screen is important – in a CU the face fills the screen, whereas in an ELS the figure's context becomes the main feature).

Movement: Mobility/immobility (as with lighting, the different categories of movement are variations within the same code).

These values are not arrived at arbitrarily. They are simply the most basic definition of the relevant variable. They are not absolute codes in that they depend on a certain cultural competence. To see a high-angled shot as a high-angled shot is an interpretation of two-dimensional imagery, and even to see a film as imagery at all demands some basic interpretative skill (as is clear from accounts of reactions to the earliest film screenings in which a shot of a train coming towards the camera led some members of the audience to panic). On the other hand, such interpretative cultural skills are not now specific to any one culture or ideological formation.

In any shot, all these cinematographic elements will enter into the meaning to some extent. In considering how they interrelate, the notion of the dominant, as used by Russian formalist writers, can be usefully adapted. The dominant is the strongest element in a shot. How it becomes the dominant can be explained by considering two features – norms and motivation.

If a particular element departs from the norms which a film uses (that is, the sub-codes employed), then this difference will make that element stand out and it thus becomes a dominant feature. In the small number of tracking shots in *Tout Va Bien* (d. Godard and Gorin, 1972) and *The Draughtsman's Contract* (d. Peter Greenaway, 1982) the movement stands out as the dominant element in these specific shots simply because both films use a norm of a completely static camera. Similarly a colour sequence in a black and white film will result in colour being the dominant in those shots which use it, such as the home movie sequence in *Raging Bull* (d. Martin Scorsese, 1980) or the coloured fish in *Rumble Fish* (d. Francis Ford Coppola, 1983).

Motivation concerns the relationship to the pro-filmic content. If an element is strongly linked to pro-filmic action, such as a tracking shot following a moving person or a high angle shot from the point of view of someone at a high window, then it will have less effect (that is, it will be less likely to be the dominant) than if the element was totally unrelated to the action. If unmotivated, it will be more likely to lead to a symbolic meaning. At several points in *Alice Doesn't Live Here Anymore* (d. Scorsese, 1974) the heroine is seen singing and playing at a piano while the camera circles around her. The lack of any character movement pushes the meaning of the tracking shot towards the symbolic. By lacking any obvious motivation, it becomes the dominant element of the shot. Features which are both unmotivated and exceptions to norms will be particularly noticeable, that is, will be very strong dominants.

It is worth noting the variety of ways in which elements may combine. One shot might contain high-key lighting (unmotivated but a norm), medium distance (motivated by distance from the hero, and also a norm), a slight low angle (unmotivated, but only a slight departure from a motivated norm) and a track around the character (unmotivated and an exception to the film's norms). In this case the movement will be the dominant. However, if the shot had been static, then the angle might have been the dominant. The same shot in a different film from a different period (using different sub-codes) might have a very different dominant.

It will be clear now that the first step in describing the cinematographic system in any film is to describe the norms of the film's style. The problem then arises as to how these are to be indicated. The majority of shots in both *The Adventures of Robin Hood* (1938) and *The Draughtsman's Contract* are static, yet clearly camera movement plays a very different role in each film. This difference can be described in part by noting when and how camera movement occurs. However, to make the difference clearer, the percentage of static shots will be a useful statistic. In the earlier film, slightly more than 73 per cent of the shots are from a static camera, whereas in Greenaway's film the percentage is over 95.

It is important to note the limitations of this kind of statistical shot analysis, especially in the face of Barry Salt's dependence on it. Salt writes of his collection of average shot lengths that it 'allows many interesting and important conclusions to be drawn with a fair degree of certainty'.[20] The first point to note is that to say that a film has 73 per cent of its shots taken from a static camera is not to say that 73 per cent of the film is shot from a static camera. The average length of the static shots may be half the average length of the moving shots. Also, many shots may include several elements – a pan, a track and a tilt, as well as moments when the camera is static. And, of course, the notion of the dominant implies that an element may be very rare but – for that very reason – may be very noticeable. A final point is that films frequently exist in different versions. Expressing stylistic features as percentages may help to iron out such differences, but the figures cannot be treated as sacrosanct. Statistical shot analysis is a very weak tool. It does give some idea of norms but it can only act as a background for more detailed analysis.

CINEMATOGRAPHIC STRUCTURING

One aspect of signification which statistical shot analysis cannot deal with is the structuring of cinematographic elements, a structuring which is crucial for the generation of meaning. Structures are created by repetition, contrast and variation. A scene might be structured by the variation of camera distance, beginning with a long shot, moving in to a series of medium shots and climaxing with close-ups. Similarly, a scene might be structured by movement, beginning with a crane shot and then continuing with a series of static shots, punctuated occasionally by pans. If such structures become very common then they may become the expected format and deviations from this format may be experienced as being in some way unsatisfactory by the audience.

An indication of the significance of this structuring can be gained by considering three contrasting types of structure. The commonest traditional form can be called the arc form. At its simplest this is an A-B-C-B-A structure. When camera distances are involved, this might result in a structure such as LS–MS–CU–MS–LS, with the long shot being an establishing shot, repeated at the end. Usually this structure will be distorted to push the CU into the latter part of the scene, giving a structure such as LS–MS–MS–MS–MS–MS–CU–CU–MS–LS. The crucial point about this, apart from providing a ready-made formula for the film-makers, is that it offers a modulated pattern of change, without any abrupt moves, and that it simply parallels the emotional intensity of the dialogue. It serves to underline the dramatic structure of the action.

A second type of structure is the alternating structure which makes use of a simple repetition and contrast pattern, that is, A-B-A-B-A-B-A-B. Thus a dialogue scene might alternate between a low-angle MS and a high-angle MCU. A chase scene might alternate between fast tracking shots and static ELSs. When this structure is clearly linked to content (as in a chase scene) then it will be highly motivated. This will tend to make it a very predictable structure. If, on the other hand, it is not motivated by the action, then it will become a much more noticeable feature. It may lead towards a symbolic interpretation and a consciousness of the film-maker's manipulation of his or her material. In this way such unmotivated structures will frequently suggest a tension by being unpredictable yet clearly formalised.

A third type of structure is the sequence-shot with virtually no internal patterning (clearly a sequence-shot could have an internal pattern by, for example, alternating regularly between MS and CU). A famous example is the kitchen scene in *The Magnificent Ambersons* (d. Orson Welles, 1942) in which the camera is static for almost the whole scene, and there is very little character movement. Such scenes are essentially unpredictable, since at no point does it become obvious what will happen next.

Structures of these kinds are most easily described (and noticed) in relation to camera distance, but the examples mentioned should indicate that any cinematographic element can be structured, even if the final effect is hardly noticeable. In addition, it should be clear that the possibility exists for complex structures in which the different elements do not work in parallel but in conflict. This should emphasise that structuring is not as simple as it may at first seem, particularly when it is seen that as well as the structuring of individual scenes, the structuring of films overall can be studied.

It remains to make explicit the ideological significance of these structures. The general principle is that the more predictable a structure, the more reassuring it becomes. Familiarity becomes a form of security. Unfamiliar or disturbing content can be rendered less strange and less dangerous by being put into such structures. Jeffrey Paine has linked this simplicity of structure with the simplicity of content found in Hollywood films of the thirties (while falling into the familiar trap of greatly oversimplifying the consistency of both the content and the style of films of this period). 'The space and time then – or the glance of the camera and the rhythm of editing – so enclosed the simplified experiences that they demanded little additional comment.'[21] This even becomes an argument in favour of the simplified content. 'The technology reinforces the content and makes an unobtrusive but decisive argument for its plausibility.'[22] The structuring effect can be compared with the effect of generic conventions. Steve Neale has described film genre as 'a coherent and systematic set of expectations' which become 'a means of containing the possibilities of reading'.[23] The analogy would be that cinematographic structuring generates a 'systematic set of expectations' in conventional Hollywood filmmaking and this can then be seen as 'containing the possibilities of reading'. It may be a less obvious kind of structuring than that involved in generic convention and this may consequently weaken its force, yet it is clearly a part of cinematic style.

CINEMATOGRAPHY AS A SYSTEM OF SUBJECTIVITY

If, as noted at the beginning of this chapter, subjectivity is an important element in the operation of ideology, then it is obviously going to be of great importance in the analysis of film. It seems equally clear that, however subjectivity functions in cinema, it must be closely connected with the use of the camera. Through the camera a viewing position is constructed; that is, a viewing subject is implied.

Despite this, there have been few coherent attempts at describing exactly how camera style relates to ideological subjectivity. Not only that, but the whole notion has come under criticism from David Bordwell in his book *Narration in the Fiction Film*. Bordwell analyses Stephen Heath's essay 'Narrative Space'[24] where he finds four different notions of 'position' are used: (1) a physical point of view implied by any perspectival image; (2) a unified position in space, implied by a series of shots; (3) a point of view of the narrative itself; and (4) subject position, a unified 'construction of the self'.[25] Bordwell argues that these are very different uses of the word 'position', with the second, third and fourth uses being metaphorical. Once these uses are separated out, the move from the point of view of the camera in (1) to subjectivity in (4) becomes difficult to sustain. For Bordwell this is part of a larger claim in which he argues against any view of the cinema which sees film as a process of telling or showing in favour of a view which sees it as a process of cues by which the viewer constructs the narrative. A film thus becomes simply a formal structure which a viewer confronts.

Although Bordwell offers little real attack on the notion of subjectivity (and his own theory tacitly assumes an ideal viewer, not far removed from a kind of subjectivity), he does make clear the need for a coherent account. The position developed here will be to argue that the four uses of 'position' *can* cohere around a notion of subjectivity and that the concepts of interpellation and point of view show how this can be done and how it relates to cinematography. That such a coherence can be described should be evident from the step-by-step relation between the four uses of 'position'. The first describes the operation of camera point-of-view in any single shot. The viewer is positioned by this if he or she takes up the implication of the shot. This sense of 'position' is both literal (where the camera views from) and metaphorical in that the shot carries with it an implied relationship to the object within the shot. (This is analogous to the distinction between denotation and connotation.) The second sense of 'position' simply refers to what happens when a series of such individual positions are put together, again both literally and metaphorically. The famous 'Kuleshov effect', describing how a viewer attempts to unify contiguous shots, discloses the automatic seeking for unity. The attempt to create a unified position through which a series of individual points of view can be understood as, in essence, the same process. But this unified position becomes more than just a metaphorical visual construct. After all, a narrative is unfolding in front of the viewer. Thus the combination of a unified viewing position and a unified narrative results in the viewing position becoming a position *for* the narrative. It becomes the audience position by which the narrative makes sense. The step from this (a position by which a narrative becomes meaningful) to an implied subjectivity, a form of the self suggested by the text, is in fact a small one. Unity of point-of-view implies unity of the narrative, which in turn implies unity of the viewing subject. How this process works will become clearer by considering the concepts of interpellation and point-of-view and how they are involved with cinematographic style.

Cultural artefacts make an address to their audience and, as it were, attempt to move them into a specific position – the position (that is, the conceptual position) at which the address is focused. To some extent all uses of the camera do this but there are certain features of cinematography which privilege a strong form of interpellated address, those moments when the film is breaking its routine in order to call out to the audience and ask it to adopt a particular ideological location. Among the main elements which do this are moving shots, direct address, low-key lighting and extreme angled shots.

Moving shots, especially elaborate crane and tracking shots, bring the viewer into a new situation. In the thirties typically beginning a scene (or even the film as a whole), they explicitly guide him or her to the point of interest. Direct address used to be thought of as a Brechtian device for breaking ideological processes and for calling attention to the filmic apparatus. However, this is not its only use (as Jane Feuer has pointed out in relation to the Hollywood musical).[26] It can also be used very directly to speak to the viewer, to call out to him or her, or rather to call out to a specific position which the viewer is invited to adopt. This use of direct address is not as rare in traditional Hollywood films as is sometimes made out and when it does occur it is often to enforce ideological subjectivity, rather than to question it. Low-key lighting is listed here because it can be used specifically as a means of exaggerating perspective and thus exaggerating the focusing effect on the viewing subject's position (although it must be said that this is a fairly weak form of interpellation, particularly since low-key lighting can have very different effects as part of the signifying system). Finally, extreme angles can have the interpellation effect by drawing attention to themselves as a means of positioning the viewer. If an unmotivated extreme high angle shot occurs, it may have the effect of asking the viewer to seek a position in relation to the text which will explain it. Thus it makes him or her seek out the focus of ideological subjectivity.

Not every occurrence of these elements will have the strong interpellation effect and they will all have other meanings as parts of the signifying system. As always, the precise meaning must be determined from the context. These same elements may well have very different meanings in other stylistic systems, remote from Hollywood in the thirties. The strong interpellation effect is dependent on the comparative rarity of their occurrence in any specific film. It is interesting to note that all four elements tend to be played down in descriptions of the traditional Hollywood style, yet many traditional films use several of these devices at crucial moments in the text.

Point-of-view may at first seem to be a simpler device but even it can yield confusions. The most detailed study of it to date is Edward Branigan's *Point of View in the Cinema* but there it is rather arbitrarily limited to what are usually called subjective shots, that is, shots 'in which the camera assumes the position of a subject in order to show us what the subject sees'.[27] In the present discussion, point-of-view is seen rather as an inevitable aspect of all camera shots. The point-of-view of the lens is the point from which the pro-filmic content is viewed. This may not always be clear, but it will necessarily always be present. Interpellation calls the viewer into the point-of-view structure. This emphasises the viewer's activity in the process of constructing a unified point-of-view. There is clearly in the text a preferred role for the viewer, a preferred subject position, but just as the viewer has to actively adopt this position (on the promise of the pleasures which the film offers), so he or she can adopt some alternative. Subjectivity of some kind is always there (the viewer is always a viewing subject) but the preferred subjectivity of the text (and the consequent ideological position) may be rejected.

From this a very important point follows. When discussing the ideological content of texts, what is being discussed is how the text has been determined by ideology, that is, what traces, symptoms, structures in the text indicate ideological causes. Such causal links may then be supported by appeal to the economic context out of which the text grew. What is emphatically *not* being discussed is how audiences are affected or how they must necessarily react. The only way in which this latter question can even begin to be answered is by a large-scale study of audience reactions which would encompass not just immediate conscious reactions but long-scale attitudinal and behavioural changes. It seems safe to assume that the data for such a study on the 1930s audience is not available. The concept of subjectivity can be described as though it implied that any text only allowed one subject position, just as ideology can be described as though it was totally monolithic and unchangeable. Such views easily lead to assumptions about audience effect. The untenability of this view should by now be clear. Any film text will (through point-of-view and interpellation) construct a subject position but the very imperfections of this process and the often conflicting ideological forces at work will allow opportunities for alternative subjectivities, alternative readings. And of course since each person will bring his or her already existing subjectivity to the film, a multitude of complex interactions is possible.

CONCLUSION

It should now be clear that the complexities of the two systems of cinematography outlined above may come into conflict with each other. A long tracking shot may introduce instability, interpellate strongly and also be a minor structuring element. It therefore becomes important to indicate how these systems work together. Since there is every possibility that they will conflict, the question becomes one of deciding their relative strengths, that is, their relative importance as indicators of ideological forces.

A hint can be found by considering the strength of the conventions governing the traditional cinematographic systems. The comparative rarity of direct address (outside of certain well-defined contexts) is an indication of the strength of the system of subjectivity (since certain kinds of direct address acknowledge the presence of the viewer and thus undermine the point-of-view structure). At the other extreme the weakness of the conventions against the unstructured sequence-shot indicate that structuring is a much weaker part of the system. Somewhere between lies signification, allowing rather more leeway than the system of subjectivity.

This hierarchy can be confirmed by considering the functions of these elements in the traditional Hollywood style. Structuring is used merely to establish familiarity and predictability. Signification, on the other hand, connotes a set of values and is thus more specific in its function. Finally, the system of subjectivity aims at nothing less than constructing a unified psyche to occupy a unified viewing position. How these systems work in practice will be seen in the following chapters, particularly in the more detailed analyses.

3 From Silent to Sound: *All Quiet on the Western Front*

To investigate the cinematographic style of the early thirties, it is necessary to begin at the transition from the silent film to the sound film. By the mid-twenties the silent film in Hollywood had reached a complexity of style and expression not always appreciated today when for many people silent film is synonymous with comedy. Such films as *The Crowd*, *The Big Parade*, *The Wind*, *Flesh and the Devil*, *Sunrise*, *The Iron Horse* and *The Salvation Hunters* show a confidence in style and narration which was quickly lost as the transition to sound began. In order to set a market for later developments, *All Quiet on the Western Front* will serve as an indicator of this transition, as a film which is poised between the style of the late silent film and that of the early sound film. The detailed analysis of its cinematographic style and its relation to ideological forces should be an effective prologue to the changes which were to come.

THE MAKING OF *ALL QUIET ON THE WESTERN FRONT*

In 1929 Carl Laemmle appointed his son, Carl Jr, head of production at Universal and immediately an attempt was begun to push the studio into Hollywood's upper rank by a programme of quality production aimed at gaining prestige as well as profit.[1] One of Carl Jr's first projects was *All Quiet on the Western Front*. The original German edition of Erich Maria Remarque's novel about the First World War had appeared in January 1929 and A. W. Wheen's English translation was published later that same year.[2] It had been a commercial success although its anti-war theme was enough to get it publicly burnt by the Nazis in 1933.[3]

Herbert Brenon was originally asked to direct but was apparently rejected when he demanded too high a fee and so was replaced by Lewis Milestone.[4] The film was made during 1929, script preparation taking ten weeks and shooting taking seventeen. This was the year in which the major Hollywood conversion to sound took place and *All Quiet on the Western Front* was originally planned as a silent film but was then shot in two versions, silent and sound. The first script was written by Maxwell Anderson, re-written by Del Andrews with Milestone's suggestions, and finally revised by George Abbott. During the shooting of the film, Laemmle appears to have been content to let Milestone have control on the set. Other important contributors to the production were cinematographer Arthur Edeson, second unit cinematographer Tony Gaudio and Karl Freund who shot the closing scene. Also involved in the film was George Cukor, employed as dialogue director (only after this was Cukor given his first directing assignment).[5]

Laemmle's desire for prestige production meant that Milestone was given the best technical apparatus available. Most importantly this meant the use of Universal's 25-foot crane. Despite Anthony Slide's claim that *All Quiet on the Western Front* was the 'first sound film to use a giant mobile crane', it had been used in (and indeed built for) the 1929 sound film *Broadway*.[6] The production values can also be seen in the fairly elaborate open-air sets and the large number of extras.

The editing of the film (by Edgar Adams and Milton Carruth) is very up-to-date for 1930, showing an awareness of Russian editing techniques. This is most obvious in the first battle scene, much of which can be included under Eisenstein's notion of 'rhythmic montage'. 'In rhythmic montage it is movement within the frame that impels the montage movement from frame to frame. Such movements within the frame may be of objects in motion, or of the spectator's eye directed along the lines of some immobile object.'[7] This is in contrast to the simpler 'metric montage' in which the editing of shots together is governed by the precise length of each shot. Rhythmic montage varies according to the content, typically speeding up as a scene reaches a climax. The classic example of this kind of editing is the Odessa steps sequence from *Battleship Potemkin*, but the battle scene in *All Quiet on the Western Front*, although a less ambitious example, is still an effective one. A track along the top of the defenders' trench is intercut with static views of the attackers, and then a track along the line of attackers is intercut with static shots of defending guns. The editing of the second of these sequences is faster than the first and shows the attackers reaching the defenders' trench. The contrast between camera movement, movement within the frame and static shots sets up a rhythm which is seldom found in traditional Hollywood films.

It is also worth noting that the soundtrack is remarkbly complex for 1929–30 in its use of sound effects and contrasting silences. The closing 'butterfly' sequence illustrates this. The whole scene contains 14 shots (including the final shot after the death of the hero, Paul, in which an earlier shot of the boys looking back as they march is superimposed over a view of a field cemetery). For the first 12 shots the sound of a distant bombardment is combined with the plaintive sound of a mouth organ. In the thirteenth shot there is the sound of a sniper's rifle followed by absolute silence as Paul's hand makes its last movements. The silence continues over the final shot of the boys marching. Such a soundtrack can seem fairly simple today, but its complexity is apparent when compared with other early sound films.

Although costing $1 250 000[8] – a huge sum for any studio in 1929 – the film was a financial success. Karl Thiede gives the domestic box-office gross at $1 500 000, and the same figure for the foreign gross.[9] Henry Forman, writing in 1933, lists it as one of the eight most successful films 'in recent years'.[10] It was also a critical success, winning Academy Awards for Best Picture and Best Director, as well as receiving nominations for cinematography and script. The *New York Times* listed it as the second best film of 1930.[11]

These facts demonstrate two points. Firstly, the film was at the forefront of Hollywood technique when it was premiered in April 1930. Secondly, it was an undoubted success, both commercially and critically. These points are important in that they show that in 1930 the film's cinematographic style was quite acceptable, both to audiences and to other filmmakers.

THE SCRIPT

The story of the film can be told fairly briefly. It begins in a German town in 1914. Seven schoolboys are persuaded by their schoolmaster, Kantorek, to volunteer for the army. They are trained in barracks by the sadistic Sergeant Himmelstoss, formerly a postman in their hometown, and are then sent to the front where they join a group of soldiers led by the veteran Katczinsky. The remainder of the film shows the gradual killing or crippling of the seven boys, ending with the death of their leader, Paul, shortly after the death of Katczinsky. Episodes include a visit to a hospital to see one of the boys, Kemmerich, after the amputation of his leg, a meeting with three French farm girls, Paul's wounding and subsequent stay in hospital, and his unhappy return home on leave.

Despite all the re-writing which the script underwent, the film stays remarkably close to the novel. As would be expected in any adaptation from novel to film, some scenes are dropped. In addition the order of events is changed slightly, although this is of minor importance since the novel has a fairly episodic structure. However, most of the scenes in the film occur at some point in the novel. Although the written version begins as the boys reach the front, the film's early scenes, showing their recruitment and training, are described in the later chapters as the narrator, Paul, reminisces. Much of the dialogue of the film is lifted intact from Wheen's translation. Even the ending of the film is not as extreme a variation of the novel's ending as has been suggested. Kingsley Canham writes that both Milestone and Laemmle 'disliked the original ending of the book in which Paul dies heroically, but neither could suggest an alternative acceptable to both parties, until Karl Freund, the famous cameraman, put forward the idea of the hand stretching out toward the butterfly'.[12] Paul is seen looking at a butterfly, an enemy sniper takes aim, Paul's hand is seen nearing the butterfly, there is the sound of a shot and his hand jerks back and slowly opens up. Although this is not in the book, it does not differ greatly in mood from the original ending. The final chapter is short and, like the rest of the book, is written in the first person. It is a meditation on the imminent return of peace and refers to the appearance of trees and berries. The image of a butterfly matches well with such a mood.

Following this meditation the chapter ends with two short paragraphs in the third person, describing his death on a day when the front was officially described as 'all quiet'. No further details are given of the circumstances. Thus not only is this death futile rather than heroic, but the ending in the film is perfectly consistent with that in the book. Laemmle and Milestone probably rejected the original ending for cinematic reasons. A visual treatment needs to show a specific incident.

CAMERA MOVEMENT

Movement of the camera is one of the most obvious distinguishing features of *All Quiet on the Western Front* and yet, in terms of overall figures, the film has a fairly normal percentage of static camera shots for a Hollywood film of its period. A sample of two randomly selected 100-shot sequences (such a sample will give a result very close to the percentages to be found in the whole film) gives 74.5 per cent of shots from a totally motionless camera. However, the remaining percentages vary in an important way from what was very soon to become the norm. 16 per cent are travelling shots (either crane or tracked), 2.5 per cent are pans or tilts, and 7 per cent contain very small reframing movements. Thus the film contains much lower numbers of reframing and panning shots than of travelling shots. In addition to this, the travelling shots are frequently very noticeable and so give rise to the impression of a film made with a fairly mobile camera.

Camera movements can, of course, be used in a number of different ways and so it is important to identify the precise uses in any given film. In *All Quiet on the Western Front* the travelling shots occur as follows.

1. Opening shot: track out through a house door to a street.
2. Classroom scene: begins with a crane back from the street through the classroom window and ends with a crane back through the window to a high angle.
3. Two parallel tracking shots accompany the boys as they march to the barracks.
4. Barracks sequence: this contains fifteen travelling shots, of which ten are of the recruits on the parade ground. The others are either in the barrack room (the first three crane shots) or in the closing scene when the boys ambush Himmelstoss (two track shots of approaching feet).
5. Two high-angle crane shots of the boys on their first night patrol at the front.
6. First battle sequence: 35 travelling shots (some quite fast) in a six-and-a-half minute sequence of 143 shots.
7. First hospital scene: crane back on entry into the hospital and a move forwards with Paul at the end.
8. Three tracking shots of soldiers marching and falling as Kemmerich's boots are passed from one soldier to another.
9. Second battle sequence: two travelling shots in the first attacks.
10. The scene with the French girls: three travelling shots at the river as the soldiers first meet the girls.

11 Travelling shot with marchers.
12 Second hospital scene: track up the ward at the beginning of the scene, track with Paul as he is wheeled out for his operation and track with him as he returns.
13 Track shot with Paul as he walks along the street in his home town at the beginning of his leave.
14 Crane into classroom (as in item 2 above) when Paul returns to his school.
15 Track with Paul at the beginning of his return to his company.
16 Three track shots as Paul and Katczinsky walk together just before the latter is killed.

This list makes clear the functions of the travelling shot in *All Quiet on the Western Front.* Camera movement is used to reflect character movement within scenes and to introduce and end scenes. In addition to this, these movements can be seen as reflecting the progression of the narrative. This progression is expressed in spatial terms. The recruits gradually move to the front and then through a number of locations – hospitals, battlefields, home. This spatial movement parallels the boys' emotional progression from optimism, through horror and disillusion, to despair. The making of a stylistic feature out of camera movement is thus closely tied to the narrative process. Movement acts both as a spur to the action and as a metaphor for the film's thematic progression.

At first it may seem odd that this use of camera mobility is linked to a low frequency of panning and reframing shots but this oddity vanishes when the film is put in its historical context. The technique of *All Quiet on the Western Front* is rooted in that of the late silent film. The extended use of pans and reframes would only develop later in the early years of the sound era. The technique of Hollywood films of the 1920s can be seen in three very different examples. *The Crowd* (d. King Vidor, 1928) contains 38 travelling shots and only 28 shots which can be classed as pans, tilts or reframes. Buster Keaton's *Sherlock Junior* (1924) contains 28 travelling shots (23 of them in the final chase sequence) and 14 pans, tilts or reframes (nine in the final chase). *The Phantom of the Opera* (d. Rupert Julian, 1925) contains no travelling shots (discounting the 11 shots of Christine and Erik in a coach during the final chase in which there is no visible background movement) and only three pans (one very minor), two tilts and two shots combining pan and tilt. Every other shot in the film is from a static camera. Kristin Thompson has described the gradual development of camera movement before the 1920s, noting influences as various as outdoor settings (which broke away from the theatreseat analogy, Italian prewar epics (most notably *Cabiria* in 1914) and the need to show off large sets (most famously in *Intolerance*).[13] However, by the 1920s these developments had settled down.

> Tracking, panning, and reframing movements remained in occasional use into the twenties. They were relatively infrequent, however. Most films show such extensive planning that the mobile frame is not necessary. . . . On the whole, aside from tracking shots following chase scenes and the like, camera movement

was a relatively minor part of Hollywood's stylistic repertory until late in the silent period.[14]

By 'late' is meant from 1926 onwards and the movement which became increasingly popular at this time was not panning and reframing but more extensive camera tracking, under the influence of German films (Thompson singles out the 1925 film *Variety* as being particularly influential).[15]

Looked at in this context, *All Quiet on the Western Front* can be seen as a transitional film between the late silent style of elaborate tracking movement with little reframing and the early sound style with frequent panning and reframing.

CAMERA DISTANCE

Camera distance in *All Quiet on the Western Front* is distributed fairly broadly. The sample of two 100-shot sequences gave the following percentages.

Extreme close-up	0
Close-up	6
Medium close-up	15.5
Medium shot	15.5
Medium long shot	22.5
Long shot	26
Extreme long shot	14.5

Two facts stand out here – the largest category is the long shot, and the full range of the middle distances (MLS to MCU) is well-represented. The predominance of the medium shot which will emerge during the thirties is not seen here. The one point in the film at which CUs and MCUs play a dominant role is in the early classroom scene. This identifies the boys and emphasises their important role – they will carry the viewer through the narrative and their reactions will be presented as a guide for the viewer.

When individual scenes are examined, they are found to have been built up by repetition, rather than by a progression inwards (such as from ELS/LS to MLS/MS to MCU/CU). The scene in which the boys first arrive at a railway station near the front is an example of this structuring by repetition. The 26 shots in the scene are arranged as follows.

1	ELS	8	MS	15	MS	22	ELS
2	MLS	9	LS	16	MS	23	LS
3	LS	10	MS	17	MS	24	MS
4	MLS	11	LS	18	ELS	25	LS
5	LS	12	MS	19	MS	26	ELS
6	MS	13	ELS	20	MS		
7	LS	14	LS	21	MS		

The scene is framed by a repeated ELS (shots 1, 13, 18, 22 and 26) which is a high-angle shot of the whole area, noticeably different from the other shots due to its being framed by a window. After the first shot there is a sequence alternating between MS/MLS shots of soldiers at a train (shots 2, 4, 6, 8, 10 and 12), and low-angle shots of horses and men in LS (shots 3, 5, 7, 9 and 11). Between shots 13 and 18 there is a LS of soldiers diving to the ground and then three MSs of individual soldiers. Between shots 18 and 22 there are again three MSs of individual soldiers. The final segment, between shots 22 and 26, consists of an earlier set-up (shot 14, soldiers in LS) appearing twice in shots 23 and 25, enclosing a MS of soldiers (a variant of shot 2 and its repetitions). Thus the first part of the sequence is based on alternating distances and the second part on units of three shots.

The first scene in the dug-out is constructed in a similar way. The 16 shots are arranged as follows.

1	ELS	5	CU	9	CU	13	MCU
2	LS/MLS	6	MLS	10	CU	14	MLS
3	CU	7	ELS	11	LS/MLS	15	MCU
4	LS/MLS	8	CU	12	MS/MLS	16	LS/MLS

Shot 1 is a shot of explosions on the battlefield. There is then a general view of the dug-out with some soldiers in LS and some in MLS. This is repeated in shots 11 and 16. Each of these is above average length – shot 2 taking 26 seconds, shot 11 taking 20 seconds and shot 16 taking 25 seconds. A related set-up is shot 6 which is a low angle of the dug-out with the soldiers in MLS. This lasts for 44 seconds. Within this frame there are repetitions of closer shots – the three CUs in shots 8, 9 and 10, and the alternating sequence of shots 12 to 15. Shots 3 to 5 consist of a repeated CU on a rat enclosing a brief repetition of shot 2. The remaining shot (7) is a repetition of shot 1 showing explosions outside the dug-out.

Although based on repetition, this structure is not as far from the common 'arc' structure as might appear at first sight. Apart from the first two CUs of the rat, CUs and MCUs are used as reaction shots. What are missing, in terms of the arc format, are the middle-range shots which would smooth the way to the close shots. It is also worth noting that when Paul starts shouting, at the end of shot 6, there is no move into a close shot to concentrate attention on his emotional state. Instead the camera remains at MLS until his outburst is over.

It is clear then that both in its choice of camera distance and in its structuring of distances within scenes, *All Quiet on the Western Front* differs from what was soon to become the norm in the Hollywood of the thirties.

CAMERA ANGLE

The most remarkable feature of the camera angles in *All Quiet on the Western Front* is the number of high-angle shots. In the sample sequences the camera was parallel to the ground in 50 per cent of the shots, placed at a high angle in 43 per cent and at a low angle in 7 per cent. The vast majority of these angled shots could be described as motivated; for example, shots looking down into a trench from the height of someone standing at normal ground level or shots looking down at someone sitting on a chair. The sample gave only 2.5 per cent of shots in which the angle was totally unmotivated by character position. The next step is to see how the large number of motivated angles and the small number of unmotivated angles are used.

The motivated angles by their very nature are unobtrusive. However, they are used fairly consistently to put the soldiers in a physically inferior position to the viewer and thus to emphasise their vulnerability and helplessness. This accounts for the fact that the high angles far outnumber the low angles. When low angles do occur they are used to get the same effect by showing us the soldiers' point of view rather than the soldiers themselves. A clear example of this occurs in the second battle scene when Paul shelters in a crater and there is a low-angle shot of soldiers jumping over him.

The unmotivated angles are reserved for more occasional emphasis. The opening shot of the barracks sequence shows the barracks gate. This is followed by an extreme high-angled ELS of the drill-ground and the surrounding buildings. This is a fairly conventional use of the high angle, showing the overall geography of the ensuing scenes. However, it is higher and more distant than might have been expected if the purpose was only to establish spatial location. This excess adds a metaphorical level to the shot, emphasising the dominance of the institution (and the war in general) over the raw recruits.

A near-overhead shot occurs at the end of the barracks sequence. The boys take their revenge on their drill-sergeant Himmelstoss by ambushing him as he returns from an evening's drinking. They bundle him up in a sheet and drop him in a large puddle of muddy water. As he is dropped in there is a near-overhead shot, very clearly showing him at his weakest. In addition, since the boys were over his head in a tree as they ambushed him, there is a weak motivation to the angle – as though the viewer were left in the tree looking down after the boys had jumped on Himmelstoss.

Another fairly conventional use of the unmotivated high-angle shot occurs in both the night-patrol sequence and during the first battle sequence. This is a high-angle travelling crane shot, moving parallel to the soldiers advancing over the battlefield. Again the soldiers are shown at their most vulnerable, becoming merely a pattern on the battlefield. (This kind of shot has of course since become something of a cliché in war films.)

Finally a more unusual high-angle shot can be found in the scene in the barrack room when the boys discover that Himmelstoss, their local postman, has become their sergeant. A travelling shot follows Himmelstoss as he walks between two rows of recruits. The high angle of the camera looks down over the tops and backs of one line of soldiers, showing Himmelstoss in MS, with the second line of soldiers facing the camera in MLS. Himmelstoss is in the centre of the frame and the faces of the further line can be made out beyond him. Here the high angle produces quite a complex effect. The importance of Himmelstoss is not in doubt – he dominates the frame (the camera moves with him past the stationary recruits) – but yet the downward look reduces his authority in the literal way of making him appear small. This is a good example of the way in which tension can be introduced into a shot by a particular combination of elements (and of course it reflects a tension in the scene portrayed).

Broadly speaking then, camera angle is used to reflect the relationships between characters. The frequent high angles, whether motivated or not, reflect in general terms on the situation of the boys.

LIGHTING

There are not many strong uses of shadow in *All Quiet on the Western Front*. The immediate reason for this can be related to a technical point. Barry Salt notes how incandescent tungsten lighting came to be used in Hollywood first in 1926, when it was considered to be better for panchromatic film stock, but then its use was increased with the introduction of sound since the old style of arc lighting made a hum which was picked up by the microphones.[16] By 1931 quieter arc lights were available and gradually regained their popularity (although Salt notes that even in 1931 80 per cent of film lighting was from incandescent lights).[17]

The importance of this historical detail is that incandescent lights gave a softer effect, as Salt notes. 'In general the light from tungsten-source lighting units was slightly softer than the equivalent unit with arc source, and when used for figure lighting they produced attractive soft-edged shadows on the face.'[18] This type of soft-edged low-contrast lighting is apparent in many scenes in *All Quiet on the Western Front*. The scenes with stronger and sharper shadows are scenes without synchronised sound, in which the hum of arc lights would not cause a problem. A typical example is Katczinsky's first appearance, in which he is seen stealing a carcase from an army butcher's truck. A strong arc source from the right of the frame produces a very dark, hard-edged shadow on the truck and a strong shadow on Katczinsky's face. Such scenes, however, are exceptions. Generally the lighting varies from fairly evenly-lit scenes with strong fill and back lighting (such as the visit to Kemmerich in hospital) to the darker scenes in the dug-out in which fill and back lighting are reduced to a fairly low level, leaving the softer incandescent lights as the key lights.

The level of lighting in *All Quiet on the Western Front* reflects the predominant mood of each scene. The darkest scene in the film is the night patrol when the recruits first see action and experience the terror of being under artillery fire. When Behm is killed in this scene the frame is almost totally dark. Darkness is thus linked to danger and isolation, with the former predominant. An example in which isolation becomes the principal connotation of darkness occurs in a scene towards the end of the film. At the end of his leave, finding himself alienated from his family, Paul decides to return to the front earlier than is necessary. He speaks to his mother in a three-shot scene. The first shot lasts for 112 seconds and shows Paul lying on his bed in MS with his mother sitting at his bedside, also in MS. Together they make up a traditional 'Madonna and Child' composition. The background is dark, serving to isolate them from any surroundings. They are lit principally by a key light from almost 90° to the right of the camera. There is very little fill lighting and so shadows are produced on both faces, the mother having darkened eyes and Paul having light shade over all his face. The second shot of the scene is a 15-second shot in which the camera is further back, showing Paul in LS, as his mother gets up and goes out the door at the back of the room. The background is still very dark, although some lighting is used (motivated by a bedlamp and by a lit room beyond the door) to pick out enough details for the layout of the room to be visible. The final shot of the scene goes back to Paul who is seen in MCU with lighting as in the first shot; that is, no lighting on the background and reduced fill so as to leave shadows on his face, notably on his eyes (at one point giving a skull-like effect). This shot last for 12 seconds. The effect is to move from the shared isolation of the two characters to the more total isolation of Paul. The eye shadows further emphasise this aspect by isolating Paul from the viewer's gaze – with his eyes not visible, his emotions can only be guessed at.

Also worth noting is the strongly directional lighting at the beginning and end of the night-patrol scene. There is an effect close to what would later become known as 'Nuremburg' lighting – strong parallel rays dividing the frame diagonally. In this scene they are meant to represent moonlight but their divergence from the lighting styles elsewhere in the film draws attention to them as a metaphorical device. A slightly spiritual or religious effect is given (rays through a church window, lights from the heavens picking out individuals) and it serves to mark the night patrol as a rite of passage as the recruits get their first experience of war.

When considered as a structural device, it can be seen that lighting is used to demarcate the episodic structure of the film. The changing moods of the different episodes are reflected in the dominant tones of each scene. This scheme can be tabulated as follows.

	Scene	*Lighting tones*
1	Introduction and classroom scene	medium
2	Barracks scene	medium-light
3	Himmelstoss's ambush	dark
4	Arrival at the Front	medium
5	Night patrol	very dark
6	1st dug-out scene	dark
7	1st battle scene	medium-dark
8	Meal	medium-light

9	1st hospital scene	light
10	2nd dug-out scene	medium-dark
11	2nd battle scene	medium
12	Paul in the crater	dark
13	The French girls	medium-light
14	2nd hospital scene	light
15	Paul's leave	medium-dark
16	Katczinsky's death	medium-light
17	Paul's death	medium-dark

This is, of course, only a very approximate guide to the predominant tones but it is enough to make clear the use of contrasting levels of light for contrasting scenes. It is worth noting how the lighting level is tied to the reactions of recognised individuals. The battle scenes are not particularly dark – there the danger is very general and comparatively few of the faces will be recognised by the viewer. On the other hand, the night-patrol scene and the scene in which Paul is trapped in a shell crater with a dead French soldier – scenes in which the participants are known and their faces can be seen – are among the darkest in the film. This reflects the film's individualised approach – the concern is with the fate of one identifiable group of men, not with the causes and progress of the war in general.

SIGNIFICATION

Having identified the film's stylistic system (the cinematographic sub-codes in use), the next step is to describe how this system as a whole signifies. Taking camera movement first, it is evident that the mobility of the travelling shots is closely related to character movement or to narrative function. The former category is unambiguous, the latter slightly less so. Yet even these functional movements (for example, at the beginning and ending of scenes) are fairly closely defined. The meaning of the opening shot in which the camera moves through a door, or the shot linking the street scene to the classroom scene, is not in doubt. Such shots have the function of moving the plot forwards, taking the viewer into the next stage of the narrative. Thus their ambiguity is very limited. Lack of corresponding character movement frees the meaning from a very close relationship to the action but the limited use of these camera movements tends to stop the meaning from becoming too independent of the plot.

There are two other important points to mention concerning the sub-code of movement. The first is that camera movement is mostly motivated by the movement of people. This is not as obvious a point as it might sound. A film might have a lot of motivated movement which is centred on animals or even inanimate objects (for example, a flowing river or moving clouds). Movement tied to human characters is, of course, the dominant form of camera movement in Hollywood films but this should not obscure the fact that there are alternatives. The second point to mention is that, as noted earlier, almost three-quarters of the shots in the sample from *All Quiet on the Western Front* were filmed from a static camera. Even in a film such as this, with many travelling shots and some spectacular crane shots, the dominant set-up is still completely immobile.

The sub-code of camera movement, then, is based on a static camera but with movement used in fairly limited, unambiguous ways, mainly to follow human characters. Taken in isolation this suggests a particular view of personality, combining belief in the autonomy of the individual with belief in the ultimate value of the individual. These are the answers to the question as to why individuals should command character movement (and remember that when the camera is not moving, it is frequently because the characters are not moving). The total dominance of male characters in the film (they tend to command the camera even on the few occasions when female characters are present) suggests that this valuation of personality is gender-specific. This is all set in the stable context provided by the dominating static camera. What movement there is does not disrupt the strength of the context. It may seem odd to talk of a strong and stable context in a film which portrays the destruction of society and which concentrates on the changes which occur in a group of boys but the point is that the value system does not change. Whether it is the boys' eagerness to save their country at the beginning, the relationship between the old soldiers and the young recruits in the middle of the film or Paul reaching out towards the beauty of the butterfly at the end, the values of individualistic, patriarchal humanism are implicit. These values are the stable background against which the narrative changes take place. In relation to this, it is worth noting that despite having no control over the war, the boys are always shown as exercising freedom of choice, a fundamental value of this system. They volunteer to join the army, they get their own back on Himmelstoss, they distance themselves from the official view of the war (in the meal scene) and Paul's death follows from his choosing to do something (reaching for the butterfly) which asserts his independence, even if only in a small way and despite it proving to be fatal.

The balance between stability and camera movement in *All Quiet on the Western Front* is worth comment. It may seem as if the more static shots a film contains, the more stable the meaning is, and therefore the more ideologically solid it becomes. That this is not the case is evident from, for example, the various films directed by Jean-Luc Godard in the decade from 1967. The link between camera movement and a particular view of personality should make clear why this is so. A totally static camera (in an era in which camera movement is technologically viable) signifies a down-grading of the importance of human will and action. It is the combination of a static camera with some character-led movement which results in the most appropriate mixture of values for an ideology which asserts human will as the only motivating force of human thought and action.

Camera distance in *All Quiet on the Western Front* serves to avoid too close an identification with one character or, at the other extreme, too strong an impact from the environment. The boys are identified as a group, with Paul only gradually emerging as the principal character and only in the later stages of the film coming to dominate the narrative. The first half of the film emphasises the group and the environment but without letting the latter dominate completely. The camera distance serves a narrative purpose which is also ideological, propping up the view of human nature already described, while keeping the group as a whole in sight. It is the group which camera distance signifies as important. When close-ups are used, they tend to be used in groups, as in the dug-out scene already discussed or in the closing shots of the first battle scene when a bottle is passed round the soldiers in the trench.

The sub-code of camera angle clearly puts the boys in a vulnerable position. The viewer is superior, not only in the security of his or her position but also, of course in knowledge. The outcome of the First World War is presumed to be known by the viewer and thus the boys' initial optimism is known to be misplaced. The force of the film is heavily dependent on this foreknowledge. The viewer looks down on the boys, not just from a position above the trenches and the drill-ground, but also from a position of omniscience. This superiority, then, will tend to lead towards pitying the boys, but pity immediately implicates the viewer in a particular moral and therefore ideological view of the film.

While emphasising the unusually large number of high-angle shots, the dominance of parallel-to-ground shots must not be forgotten. It is by contrast with these that the high-angle shots have their effect. The significance of this is that camera angles alternate between the high view of the boys' vulnerability and the parallel view which keeps the viewer at the same level as the boys. The occasional low angles push towards a closer identification with them. The film as a whole keeps these three tendencies in a balance which corresponds to its ideological project.

Shadow operates principally to signify the presence or absence of danger or isolation. By establishing this sub-code in the earlier part of the film (or rather, by establishing that this common sub-code is being employed), particularly in the night-patrol scene and the first dug-out scene, it can then be used in more ambiguous scenes. The lightest scenes are in the hospital. Although there is some danger here, the hospital is essentially a place of security (away from the fighting but also clearly separate from the pressures of home) and community. This indicates how closely lighting follows the narrative content. In scenes such as these, it is merely underlining this content, rather than coming into any kind of conflict with it.

Two additional points are worth noting about the lighting, both being absences. There is an absence of traditional star (or 'glamour') lighting, that is, the kind of strong top lighting that creates an aura around a head. Such lighting is most usually associated with female stars but its absence in any form in *All Quiet on the Western Front* helps the integration of the group and the setting by not singling out any one character. The other absence is the consistent use of strong shadow on faces. This absence tends to lessen ambiguity and to integrate the group by not concentrating on conflicts within individuals.

Generally, then, the sub-codes of signification work to restrict ambiguity, to underline the narrative process, to support a particular view of human nature and to put the viewer in a specific moral position. These features are not, of course, independent of each other but overlap in many ways. When taken together they coalesce around the film's ideological project which, put very simply, is to condemn war because it violates these assumptions about human nature in particular the valuation of human nature as an end in itself. There is no analysis in the film of economic, social or political causes and class differences are raised merely to be shown to be unimportant (the boys are clearly middle-class, just as the older soldiers are clearly working-class). The individual's qualities are seen as transcending any social or economic structures, and the war becomes a meaningless event with which the soldiers have to cope.

The structuring of cinematographic elements within scenes also has its ideological significance. The use of repetition gives scenes a definite structure but without the predictability which frequent use of an arc pattern would give. The structuring in *All Quiet on the Western Front* is often fairly obvious (most notable, perhaps, in the first battle scene) but that does not, in itself, make it predictable. If the strongest structuring effects are the predictable arc and alternating patterns, and the weakest (or rather, the most open) are long unpredictable sequence-shots, then the relatively complex repetition structuring in this film falls somewhere in between. As noted in Chapter 2, even very predictable structuring is a comparatively weak feature, and so here it becomes even weaker.

Not only does this structuring have little predictability, but it is also less obviously aligned with the filmic content. An arc structure can carefully guide the viewer through a scene, underlining the emotional content. In *All Quiet on the Western Front* this does not usually happen. Scenes frequently avoid using close-ups at emotional climaxes. The first dug-out scene, with the camera remaining at MLS during Paul's outburst, is typical. This refusal of too obvious a parallelism between structure and content does introduce a certain (although very limited) ambiguity.

Structuring reveals something else about a film. To structure a film in the dominant style is implicitly to categorise it as part of the dominant practice. By not using dominant structures, the makers of *All Quiet on the Western Front* place it outside that style. This, however, does not necessarily indicate ideological difference since it can be used quite effectively to give the film the status of a prestige event (as Laemmle, for one, clearly intended). Any departures from established norms simply confirm the film's status as exceptional.

It is important, however, not to exaggerate the film's differences. *All Quiet on the Western Front* belongs to a transitional period, well-recognised as such by film-makers and audiences at the time. This period generally lacks the strong structuring that later became common. A few months after this film, Universal released *King of Jazz*, a review musical featuring Paul Whiteman. Although very different from the earlier film, it was also intended as a prestige production, with extravagant production numbers and shot in two-strip Technicolor. As well as the musical numbers, *King of Jazz* includes some short comic sketches (one of which is a parody of the scene with the French farm girls in *All Quiet on the Western Front*). These sketches are mostly in long or medium long shot, with occasional inserted close-ups. They do not show the common structure of later films. Other films from 1930 show a similar structuring and it can be traced back to the style of the Hollywood late silent film.

It would appear then that the structuring in *All Quiet on the Western Front* is less unusual when put into the context of Hollywood in 1930 than it would appear if compared with the later thirties. If then, the film's difference from other films of the period is not great, the difference that there is can easily be explained by the audience in the terms of the prestige production. Once this status is granted, rather than being an ideologically unusual film (because of its stylistic features), it simply becomes a film which confirms existing norms by indicating their limits. By moving slightly beyond the conventional style, but also being celebrated as a 'quality' film, *All Quiet on the Western Front* becomes ideologically 'safe'. Its minor unconventionalities 'prove' its quality, and its quality 'proves' the acceptability of these features. Both elements 'prove' the film's ideological innocence.

SUBJECTIVITY

In discussing subjectivity, the first question must be how the viewer is addressed. The opening scenes are instructive. The clear interpellation devices of the opening track through a door to the soldiers in the street and the later crane back through a window to the classroom, address the audience very clearly, putting it in a position of superiority over the protagonists and, in effect, bringing about a partial distancing from these protagonists. This is because the audience is moved independently of them and has knowledge independently of them (after all, the street *and* the classroom are seen on the screen – the characters in the film see only one of these).

This links up with the organisation of point-of-view. Point-of-view in *All Quiet on the Western Front* can be pictured as a set of concentric rings. In the centre are the scenes organised around Paul's consciousness (for example, his return home on leave, in which only events which he witnesses are seen). The next ring contains the scenes tied to the boys as a group. The camera adopts positions which unite around the idea of another boy – the camera becomes one of the group and does not single out any individual for particular emphasis (the scenes in the barracks are examples of this). The third ring contains those scenes which are more generally associated with soldiers at large, rather than just the group of boys (for example, the battle scenes). Finally beyond this there is an outer ring consisting of those isolated shots which put the audience at a distance (metaphorically but usually also literally) from the action. The ELS through a high window which recurs in the station scene discussed earlier is an example, as are the final shot in the first classroom scene (as the camera cranes back to a high angle of the empty room and holds that position for a few seconds) and the crane shots of the battlefield. The justification for representing these point-of-view structures as concentric rings is that Paul is one of the boys, the boys are all soldiers and the soldiers are contained within the view of the larger landscape. Thus there is a varying but overlapping set of positions for the viewer in the inner three rings – that these rings are concentric shows how close the positions are. Only in the few more distant shots is this subjectivity weakened and a point-of-view more distant from the protagonists is suggested. This distancing allows for a break or at least a weakening of the binding process formed by the editing structure (there are no reverse shots at these distant points).

The subjectivity offered by the distant shots is much freer of the diegetic characters and much less defined. It begins to suggest a narrator's subject position – someone outside the diegesis who is addressing the viewer in a particular way, that is, an authorial subject. It is no accident that these shots are usually also the most obvious interpellation shots. It is the author-narrator who calls out to the audience, asking it to adopt a specific position in relation to the narrative. The fact that these shots are also crucially important in the signification system suggests how important and complex their function is.

Two final points can be made. The first is that the appearance of an authorial subject position can function, like the unconventional structuring, as another 'proof' (perhaps the ultimate one) of the film's quality. The film becomes linked, however tenuously, to the values of high art and to the seriousness of artistic (that is, authorial) statement. Thus the film's ideological unity is not threatened. The second point is a more general one. Subjectivity is not being equated simply with point-of-view. The overall subjectivity of the film contains all the variations of the point-of-view implied by the concentric rings. What is important is that these variations are related to each other (as the idea of concentricity is intended to demonstrate) and it is because of this that the subjectivity of the film is unified.

CONCLUSION

Taking the systems of signification and subjectivity together, it will be clear that *All Quiet on the Western Front* presents a fairly unified ideological structure, containing only minor ambiguities and deviations. The more overt ideological content of the film – its attitude to the war, its playing down of class difference, its assumptions about human nature and gender roles – is not, therefore, threatened by the cinematographic style.

Perhaps the most important conclusion, however, concerns the film's historical context. As has been argued, *All Quiet on the Western Front*, although differing from the norms of 1930 only in small ways, uses a style which is not that which was soon to become established as the Hollywood standard. Despite this, it was a great popular success. This is all the more surprising considering the seriousness of the subject, its distance from the American audience's own experience (both in time and, more importantly, in place) and its downbeat ending. It did not even feature a major star (although Lew Ayres, who played Paul, was already established in Hollywood, having played opposite Garbo in *The Kiss* the previous year). This should make it clear that, however the changes of style which followed in the thirties are to be explained, audience reaction alone will not be adequate. The kind of historical account which sets up a model in which stylistic innovation is developed because the audience is favourable to it, or ignored because the audience does not respond, will not be sufficient. The great difference in style which will be evident in *Dr Jekyll and Mr Hyde* must have had other causes.

4 A Crisis of Explanation: The Early Thirties

The previous chapter has indicated the stylistic norms existing in Hollywood as the transition to sound took place in 1929–30. However, as the Depression worsened and as the studios, at the same time, got used to dealing with sound, these norms began to change. By 1932 Hollywood feature films were noticeably different when compared with the output of 1930. Before describing and attempting to understand these changes, the economic and social problems which the country was undergoing need to be appreciated.

THE CRISIS OF THE EARLY THIRTIES

The economic crisis which hit the United States in the early thirties was massive and unexpected. Taking the level of unemployment, the gross national product and the rate of economic growth as the principal indicators, the situation can be easily described.[1]

Year	*Unemployed*		*GNP*	*Economic growth*
	Number	*% Workforce*		
1928	1 982 000	4.2	97.0*	0.2**
1929	1 550 000	3.2	103.1	2.3
1930	4 340 000	8.9	90.4	−0.8
1931	8 020 000	16.3	75.8	−2.2
1932	12 060 000	24.1	58.0	−4.4
1933	12 830 000	25.2	55.6	−4.0

*Gross national product is expressed in $billions by 1975 prices.
**Economic growth is calculated from a 1926 base.

These statistics show the speed at which the Depression set in. It is more difficult to find reliable figures to show the effect on Hollywood. The following figures were produced by the film industry itself and are therefore not totally dependable. However, they do show the trends of the time. The figures for weekly film attendance came from *The Film Daily*, a New York magazine, and the box-office receipts come from figures produced by the Motion Picture Association of America (although not published at the time).

Year	*Average weekly film attendance* (millions)	*Box-office receipts* (millions of dollars)
1928	65	(unavailable)
1929	80	720
1930	90	732
1931	75	719
1932	60	527
1933	60	482

These figures show clearly the novelty effect of sound, with a dramatic increase in attendance in 1929–30, but with the Depression hitting the box office in 1931–32. The following year, 1933, turned out to be the poorest year of the decade for attendance but by that time the worst of the Depression was passing. The increase in unemployment was less than in the previous three years and economic growth was not quite as bad as in 1932.

The social upheaval caused by the Depression was manifested in various ways. In some places major demonstrations took place, as William E. Leuchtenburg has described.

> As the bread lines lengthened, the mood of the country became uglier. In July, 1931, 300 unemployed men stormed the food shops of Henryetta, Oklahoma. An army of 15,000 pickets marched on Taylorville, Illinois, and stopped operations at the Christian County Mines in 1932. In Washington, D.C., 3,000 Communist 'hunger marchers' paraded. None of these demonstrations matched in importance the rebellion of American farmers. From Bucks County, Pennsylvania, to Antelope County, Nebraska, farmers banded together to prevent banks and insurance companies from foreclosing mortgages.[2]

The most famous demonstration was that of 22 000 ex-servicemen in Washington in June 1932, marching to force the government into early payment of bonuses due in 1945. The 'bonus marchers' were brutally dispersed by the army under General MacArthur.[3]

It was not only the working class who were hit by the Depression, as M. A. Jones notes.

> Large numbers of middle-class folk lost their jobs, their savings, and, worst of all, their sense of security. Tens of thousands of businesses went bankrupt. Doctors, lawyers, and architects saw their income shrink and often found themselves idle much of the time. Many college students had to abandon their studies

through lack of funds or, if they completed their courses, found themselves unemployable.[4]

The most visible signs of the crisis, apart from the breadlines, were the large numbers of vagrants (Jones estimates that 'at any given time there were perhaps as many as five million')[5] and the shanty towns (ironically called 'Hoovervilles') which arose on the outskirts of many cities.

The possibility that such economic and social upheaval might result in political upheaval was apparent to some. MacArthur justified his treatment of the bonus marchers by the threat of revolution. 'It is my opinion that had the President not acted today, had he permitted this thing to go on for 24 hours more, he would have been faced with a grave situation which would have caused a real battle. Had he let it go on another week I believe that the institutions of our Government would have been very severely threatened.'[6] Of course political upheaval did not in fact occur – the political institutions survived unscathed and there was little violence caused directly by political (as opposed to economic) issues. On the other hand, Roosevelt's New Deal government – at least in its initial programme – could count as a non-violent revolution when compared with the inactivity of Hoover's administration. And the lack of actual revolution did not diminish the fears of many at the time.

A crucial point to remember when considering people's reactions is the problem of explanation. In the oil crisis of the early seventies an explanation was at hand – the rise in oil prices – which was directly related to decisions made by a small number of nations and these decisions could be clearly linked to political factors (the Yom Kippur War of 1973). In the early thirties, no such explanation was at hand. The twenties had seemed to prove that American capitalism worked and would eventually work for anyone who wanted to take advantage of it. The bubble had burst dramatically in October 1929 with the Wall Street Crash and no explanation seemed to fit. Political leadership had no answer, with President Hoover refusing to intervene in the economy until 1932 and even then only helping banks and other large companies.[7] M. A. Jones sums up the President's approach before that intervention. 'Believing that the country's problems were more psychological than economic he issued a stream of reassuring statements, minimizing the number of unemployed and predicting that prosperity would soon return.'[8] Even after his interventions in 1932, he still refused to give help to individuals. Jones describes the attitude of the public to him at this point. 'Hoover thus became the butt of sardonic jokes, his name a synonym for misery and hardship.'[9] With no convincing explanations at hand, and a lack of confidence in political and economic leadership, ideological crisis ensued.

HOLLYWOOD IN THE DEPRESSION

In Hollywood, the early thirties really began in 1931. The previous year the studios were still consolidating the change to sound and, as *All Quiet on the Western Front* has shown, the style apparent in that year still owed a lot to the silent film. By 1931 the films were more clearly different from those of the late silent period, and by that time the Depression had begun to reach Hollywood. This early part of the decade ends during 1933 as Roosevelt is inaugurated, with the National Recovery Administration logo's appearance at the end of many films in that year signifying that the 'new deal' sense of optimism was being adopted in Hollywood.

The expected reaction in cultural products to the crisis of the Depression might have been assumed to be that the relative autonomy of the ideological superstructures would be reduced as the dominant class struggled to contain the crisis by all possible means, particularly in Hollywood since the decade before the early thirties had seen an expansion of outside investment in the film industry, thus tying it closer to the financial institutions of the country.[10] To understand what in fact happened it needs to be remembered that, as a leisure industry, Hollywood was particularly vulnerable to economic recession. When money is scarce, people will tend to cut down on leisure spending first. As a result of this, rather than a narrowing of film content, an expansion took place, in order to retain the audience. The immediate need in 1931 was to get the audience in and this was done by giving them whatever content could be shown to be popular. Only later in the decade, after the worst of the crisis was over, did retrenchment take place.

The combination of popular content and a crisis of explanation resulted in Hollywood producing many films in the early thirties in which a hero or heroine struggles against an unsympathetic environment, either successfully (the fallen woman, the hero fighting the monster, the anarchic comedian) or unsuccessfully (the gangster, the monster as hero). The theme which links a wide variety of films is that of survival. *Blonde Venus*, *Duck Soup*, *I am a Fugitive from a Chain Gang*, *Trouble in Paradise*, *Red Dust*, *The Miracle Woman*, *Arrowsmith*, *A Bill of Divorcement*, *The Old Dark House*, *Frankenstein*, *Freaks*, *A Farewell to Arms* – all are linked by an emphasis on survival in a society in which social structures are either breaking down or else are constraining and alienating the individual. This is true even when the scene is displaced to a tropical plantation (*Red Dust*), the central European countryside (Frankenstein), a remote farmhouse in Wales (*The Old Dark House*) or high society in the capitals of Europe (*Trouble in Paradise*). Even the gangster films fit into this pattern with their structure of the individual surviving against the odds (that is, against society) right up until the final scene.

CINEMATOGRAPHIC STYLE, 1931–32

In the years 1931 and 1932 camera style differed significantly from that which was to predominate later in the decade. Full high-key lighting, in which there is no discernible shadow and a very close balance between key, fill and back lights, is not a feature of the style of the early thirties. The early standard is mid-key in which fill lights are not very strong and some shadow appears. It is worth noting that the lighting units required to produce high-key, and in particular the long 'broads' which provided fill lighting, were available and being used in Hollywood in the thirties before high-key developed as a norm.[11] Clearly the units had to be in use as a necessary condition for the style but there is still a significant time lag between the technological and the stylistic developments. David Bordwell notes one factor at work here. 'After 1931, most cinematographers chose to keep the lens at full aperture, cut down the light levels, and save money on the set.'[12] If the developments in lighting are linked with the reframing style in which small camera movements are continually used to reframe so that characters are nearly always in the centre of the frame, then the core of the stylistic paradigm in the thirties is described. Other elements can be put into two categories – those which are used frequently in some types of film and occasionally in many others, and those which are used only infrequently.

In the first category can be put fairly dark lighting, with the use of strong shadows on walls and faces, and unmotivated camera angles, typically either low angles of characters in MS to MLS or high angles of a scene in LS to ELS. In the second category (devices used infrequently) can be put shots in which characters look directly at the camera, overhead shots, and extreme low-key lighting (that is, scenes in which there is very little lighting of any kind). To emphasise the selective nature of this early thirties paradigm, it is worth noticing some devices which fall outside it, such as naturalistic lighting and certain types of unmotivated camera movement (such as movement around a static subject). Concerning the structuring of sequences, there is a tendency against analytical editing in the early thirties, with the use of sequence-shots and scenes with relatively few reverse cuts (both being devices which tend toward a less predictable structure), although having said this, analytical reverse-cut editing is undoubtedly the dominant model.

If, as argued earlier, films of the early thirties, with their emphasis on social breakdown and individual survival, are seen as evidence of cultural crisis, then this should be evident at the level of cinematographic style as well. It is at this point that the signifiers of instability, ambiguity, obscurity and unpredictability become important. The generally dark tones of the lighting, with occasional plunges into low-key, signify obscurity, isolation (as the characters' relationship to their environment is, literally, hidden) and ambiguity (the consequence of obscurity). The generally freer use of camera movement (in comparison with earlier and later styles) adds a mobility and instability to the films, as well as ambiguity (particularly if the movement is not motivated by character position) and unpredictability. Low and high unmotivated camera angles again add unpredictability but, more importantly, they ask the viewer to take up unusual positions in relation to characters on the screen and thus work against the construction of stable, unified subject positions. Extremes of camera distance, particularly very close shots, work in a similar way, although in a weaker fashion, by putting the viewer at a specific distance from the characters, a distance which is significantly different from that to be expected in conventional social situations. The frequency of looks direct to the camera is particularly noteworthy. This device alters the conventional subject position by putting the viewer in the exact position of a diegetic character, thus confusing a character's subjectivity with that of the absent viewer. This foregrounds, and thus can question, the construction of subjectivity (an example of this will be seen in the following chapter). Finally, the structuring of scenes generally adds unpredictability by the use of sequence-shots, unconventional patterns and less reliance on reverse-cut editing than would become the norm a few years later.

The three films which will illustrate this style and which represent different genres, different studios, different directors and different cinematographers are *The Public Enemy*, *Susan Lenox, Her Fall and Rise*, and *The Most Dangerous Game*.

THE PUBLIC ENEMY

The Public Enemy was made at Warner Brothers, directed by William A. Wellman and photographed by Dev Jennings. It first appeared in April 1931, just three months after Warners had inaugurated the early-thirties gangster cycle with *Little Caesar* and it deals with the rise and fall of the gangster Tom Powers (played by James Cagney).

Most of the interiors in *The Public Enemy* are shot in mid-key, with light shadows appearing on background walls and some light shading on characters' faces. A slightly darker version of this is found in the scenes in clubs and speak-easies. The most notable sequence with much darker lighting is the attempted robbery in the '1915' section of the film. Tom and his friend Matt (Edward Woods) along with an accomplice, Limpy Larry, break into the Northwestern Fur Trading Company. The sequence begins in near darkness, the robbers appearing only as foreground silhouettes in five shots. As they are disturbed, the large shadow of their interrupter is seen on a wall. They are then chased into an alley where they shoot their pursuer. The killing takes place in a shot in which half the frame is in darkness, with a strip of light seen in the middle, and a greyer wall to the right. The gunshots appear as flashes in the darkness. No other scenes are quite as dark as this one, although there are others which are darker than anything which could be called mid-key. At the end of the film there is a low-key scene when Paddy Ryan (Robert O'Connor) tells Mike Powers (Donald Cook) that the Burns mob have kidnapped Tom from the hospital. The two men face each other outside the Powers' house and each is picked out by a key light. Background light is motivated by light shining through the house door. There is also very slight fill lighting.

Unmotivated camera angles appear fairly frequently in *The Public Enemy*, especially low angles looking up at characters in conversation. Examples are found in the scene showing Mike and Tom's argument before the former goes off to the war, and in the scene at the end between Mike and Paddy, mentioned in the previous paragraph. In some scenes the angles are slight. When Tom and Matt put pressure on a speakeasy owner, the camera is at bar height, looking up at the three men. More extreme angles include the road-level camera looking up at Tom as he staggers along the gutter after shooting Schemer Burns (this shot could be regarded as being motivated by the fact that Tom will fall down to the level of the camera but this does not lessen the initial impact of the low angle). Another example is in the film's closing scene when Tom's body is delivered to his home. There is a low-angle shot as Mike opens the door and then a cut to a closer and steeper angle of the body, before cutting back to the first position.

Also worth noting are some strong angled shots which are motivated by character position but still have a striking effect. The first robbery scene includes a very high-angled shot of Limpy Larry lying in the street, followed, four shots later, by an extreme low-angle shot from street-level looking up at the lamp-post down which Tom and Matt will escape. Although both these shots are strongly motivated (the first follows a shot of Tom and Matt looking down and the second is from a conventional street-level position), the strong lines of the composition and the extremity of the camera angle mark them out as exceptional. Angled shots are in fact present in the film from the very opening. After the titles, there is a sequence of documentary footage of Chicago. The series of seven shots includes five strong high angles of, for example, stockyards and street corners. These are, of course, fairly conventional documentary shots and as such have a limited expressive effect. They do, however, serve to introduce unmotivated angles into the film, indicating this as a potential part of the filmic system of *The Public Enemy*.

More unusual than the lighting or the high and low angles is the use of direct-to-camera shots. The first occurs very early in the film. In the opening ('1909') section Tom and Matt as young boys are seen being chased through a department store. In the first shot of this sequence the camera moves back quite quickly as they run directly towards it, looking into the lens. This is immediately followed by a similar shot but from a more distant and static position. Later in this opening section of the film Tom looks directly into the camera when he confronts his father. The reverse cut shows his father looking down at Tom, very nearly direct to camera, but not quite. The most extended use of these direct shots is in the scene in which Tom kills Schemer Burns. Tom waits in the rain for Burns to arrive. In a sequence of eleven shots, showing the arrival of the gangsters and Tom following them into their lair, there is one shot of Tom looking directly to camera and five in which his look is only very slightly off centre but still close enough to disturb the prevailing point-of-view patterns. Finally, the film closes with a direct shot. After Tom's body falls in the door, Mike looks directly at the low-angle camera and then walks past it. Such uses of the direct-to-camera shot were not unique. Other gangster films of this period (such as *Little Caesar* and *City Streets*) use it at climactic moments, as do films in a number of other genres.

Camera movement is not particularly unusual in this film although there is a slightly freer use of movement than would become normal later in the decade. Two examples will illustrate this. The scene in which Tom kills Putty Nose has a fairly minor camera movement but one which is unmotivated by any character movement. Tom and Matt have followed Putty Nose into his apartment. A MLS of all three is followed by a pan across to the piano as Putty Nose and Tom move towards it. Putty Nose begins playing the piano and Tom takes out his gun. The camera then pans back to Matt, still standing at the door, and moves in slightly closer to him. Gunshots are heard. The camera remains on Matt as Tom reappears and exits by the door behind Matt who pauses before following Tom out. The shot lasts for 46 seconds. Although the camera movement involved is a fairly routine pan across a room, it is very noticeable. Not only does it not follow any character, but it takes the audience away from the main point of interest and is thus fairly disruptive of the coherence of the audience's point of view, centred as it is on that point of interest.

The other moving shot of particular note is at the beginning of the film. Immediately following the seven 'documentary' shots there is a close-up of beer taps in a bar, and then a complex travelling shot lasting for just over a minute. It begins with a view of a horse and cart coming out of a brewery and pans round with the cart through 45° to where a man is seen emerging from a side street and crossing the road. The pan continues with him as he crosses the road and reaches a second man carrying six beer-tins on a pole. By now the camera has moved through 90° from where it picked up the first man. The beer-carrier is followed across a street (which is directly opposite the brewery entrance at which the shot began) and then the camera pauses, looking at the next street corner. A Salvation Army band goes past and the camera tracks down the street with it until the family entrance to a bar is reached. The band marches on but the camera stays at that point giving the first view of Tom and Matt (as young boys) as they emerge from this door.

The complexity of this shot is such that it is very difficult to retain a sense of direction at a first viewing. However, the narrative does not require that the geography of the shot is understood – the only important event is the emergence of Tom and Matt. More important is the symbolic meaning of the shot – the journey from the brewery to the man going into a bar, then to a man coming out with drinks, then the Salvation Army, and finally the boys coming out of a bar and entering upon their life of crime. The shot virtually points out the apparent argument of the film, but does so in a way which draws attention to the film's enunciation, departing from the 'invisibility' of conventional film technique. For a short period the film becomes an argument addressed to an audience, rather than a narration of a story.

In terms of scenic structure, *The Public Enemy* has some deviations from the routine patterns. The scene in which Tom and Matt are told of the plan for their first major robbery takes place in a back room at Putty Nose's club. It is a sequence-shot lasting for 78 seconds. There is a faint air of mistake about the scene, as if the director had forgotten to shoot any cut-away shots. At one point Putty Nose is speaking but only the other gang members are seen, all looking towards Putty

Nose, who is out of the frame. However, whether this sequence-shot was a consequence of the speed of production or whether it was intended, the fact remains that it was left in the final cut (it presumably would have been easy enough to shoot a few close-ups of Putty Nose after production had officially stopped).

Many other scenes in the film have only a few close-ups cut in and this gives a less predictable structure than later conventions would insist on. The following scene in which Mike and Tom talk before the former goes off to fight in the war, is an example of this.

Shot 1 MLS. Mike packing, then Tom joins him.
Shot 2 MLS. Slightly closer than the previous shot.
Shot 3 MLS, as shot 1, but with a reframing pan.
Shot 4 Begins in MLS. Pan over to the door with Mike and then back to the two-shot. Move into MCU. Back to MLS. Pan across the room with Mike.
Shot 5 CU of Tom.
Shot 6 MCU of Mike.
Shot 7 CU of Tom.
Shot 8 MCU of Mike.
Shot 9 CU of Tom.
Shot 10 LS. Mike hits Tom. Pan with Tom as he falls. Mike leaves the room. Move into MS on Tom and then tilt down to his feet as he kicks the door.

Shots 5–9 are 'over-the-shoulder' reverse-cut shots. This scene is clearly related to a conventional pattern. A MLS establishes the space of the action. A pan is used to break up the scene in the middle. There is a sequence of close shots using reverse cutting to build up to a climax and then a move back to a more distant position for the end of the scene. However, certain elements disrupt this structure. One is the lack of reverse cuts at the medium distance in the first part of the scene. Another is the length of the fourth shot – 57 seconds, out of a total length of 156 seconds for the ten shots – and the use of camera movement within this shot to vary distance. Another is the use of a single shot at the end to cover five distinct actions – Mike's punch, Tom's fall, Mike's exit, Tom's standing up, and Tom's kick. The effect of shots 4 and 10 is (at least partly) to throw into relief the closer sequence in between, making it stand out as a particular technique, rather than simply being a conventional pattern. This makes the structure of the whole scene appear as more noticeably 'crafted' and hence less predictable, that is, less 'invisible' in technique.

SUSAN LENOX, HER FALL AND RISE

Susan Lenox was made at MGM, with direction by Robert Z. Leonard and cinematography by William Daniels. Premiered in October 1931, it is notable as the only film which featured both Greta Garbo and Clark Gable. It is not so different from later norms as *The Public Enemy* but its opening sequences do contain some striking shots, particularly where the lighting is concerned. The film begins with two exterior ELSs of a winter's night, both very dark, followed by two exterior LSs as a traveller arrives at a farmhouse. These are also very dark shots. They are followed by an interior sequence which is best described as being well on the dark side of mid-key and so continues the general dark tone. The scene is the birth of Helga and the birth itself is shown by shadows on a wall. This shadow motif is then developed in a montage sequence of Helga's youth. A fade-out follows the shot of the shadow of the doctor holding up the new-born baby. The sequence then continues as follows.

Shot 1 Fade-in. Wall shadow of a small girl holding a broom.
Man's voice: 'Helga! Get to work there!'
Shot 2 Wall shadow of a young girl in front of a man's outstretched leg.
Man's voice: 'Helga! Pull off my boot!'
Shot 3 Wall shadow of a girl with an armful of sticks.
Man's voice: 'Helga! Come here!'
Shot 4 Wall shadow of a young woman washing dishes.
Flashes of lightning.
Man's voice: 'Helga!'
Shot 5 Two men at a table. Mid-key lighting. Helga walks in, her face seen for the first time.

These five shots last for 47 seconds.

Such a sequence works in several ways. It is clearly a useful way of showing the childhood of a character who is played by a famous star. It also falls into a tradition of dramatic first appearances by Garbo. However, only to see this sequence in these terms is to miss a major part of its significance. The use of shadows without a visible source, along with a low level of lighting, puts into play particular codes of signification. And in historical terms, such an extensive use of unsourced wall shadows would not be considered to be a serious option by the later mid-thirties.

Following this sequence, Helga is introduced to her intended husband, Mondstrum (Alan Hale). After her father goes to bed, Mondstrum goes up to Helga's room and attempts to rape her. The lighting in this sequence is as near to 1940s low-key lighting as was possible with the technology of the early thirties. Even some of the framing looks like classic *film noir*. The main part of the sequence is as follows.

Shot 1 Helga in her bedroom, in MLS, sitting on her bed at the right of the frame. The background is in almost total darkness. Light from the right of the frame picks out her face, the top of the bed and the wall behind her to the left of the frame. There are flashes of lightning.
Shot 2 Mondstrum comes up the stairs. The camera is at a high angle looking down the stairs towards him in MLS. There are very strong

compositional lines from the sloping ceiling, the stairs, the walls and the bannisters. Strong shadows frame his face.

Shot 3 Helga in MCU, lit by one light on the side of her face.

Shot 4 Helga's point of view of the door. The frame is almost completely dark except for small cracks of light between the planks of the door. It begins to open.

Shot 5 Reverse cut. Mondstrum's shadow appears on the wall to the left of the frame. There is some light on the side of Helga's head but her face is completely in shadow.

Shot 6 Reverse cut. Mondstrum at the door, with light behind him and his face completely in shadow.

Shot 7 Repeat of the set-up in shot 5.

Shot 8 Mondstrum in the doorway in MS. His face is visible but is not well-lit.

Shot 9 Helga in MCU, with strong light on her hair from the left of the frame. The background is totally dark. Her face is barely visible, with her eyes picked out by lights.

They then struggle and Helga runs out of the house. She takes shelter in a barn belonging to a neighbour, Rodney Spencer (Clark Gable). When he discovers her in the barn there is another shot in almost total darkness.

Nowhere else in the film is the lighting so extreme and in fact the next section in Rodney's house is the lightest part of the film. Since this section functions as the description of a paradise which the protagonists spend the rest of the film trying to recapture, the narrative function of the opposition between the dark opening sequence and the following light sequence will be clear. The rest of the film ranges from a fairly light mid-key to a fairly dark mid-key, reaching low-key only in isolated shots. One example occurs after a circus-owner has hidden Helga – now renamed Susan – in his rail compartment and has made clear what he expects in return. The scene ends with both characters in silhouette before the fade-out.

Although no other stylistic feature is as prominent as the lighting in the opening sequence, there are a number of unmotivated camera angles, particularly, once again, in this opening section, for example the first interior shots as the doctor arrives at the house are unmotivated low angles. Camera movement is not particularly noticeable. The most elaborate movement is the crane shot which opens the Paradise Café scene at the beginning of the final part of the film. A high-angle camera looks down over the balcony to the main floor of the café. It then moves down to a level just above head-height on the lower floor and moves across to the bar. This shot is simply a variation of the 'survey' shot, establishing the geography of a scene by moving through it. Also notable are the overhead shots in the montage sequence illustrating Susan's 'rise'. Overhead shots in such sequences lose some of their impact because of the context, which allows excess within certain bounds (just as such shots in musical numbers are not as disruptive as they would be in other sequences). Despite this, the shots are still unusual enough to have some dramatic impact. In terms of subjectivity, overhead shots are extremely disruptive and the conventional contexts can only partly defuse this.

Susan Lenox is clearly less unusual in style than *The Public Enemy*. In terms of scenic structure it is not unusual at all, with very predictable contrasts between distances, reverse cutting and camera movement motivated by character movement. Despite this it still shows that the stylistic paradigm of 1931 was fairly wide. This is particularly interesting since the film is the only one of these early-thirties films to have been made at MGM, usually regarded as the most conservative of the major studios. That it was not an oddity, even at MGM, can be seen by considering that another Garbo film of this period – *Grand Hotel*, released in April 1931 – includes overhead shots and occasional complex movements, as well as some very dark set-ups.

THE MOST DANGEROUS GAME

The Most Dangerous Game (still known in Britain as *The Hounds of Zaroff*) was made at RKO by much the same team as went on to make *King Kong* a few months later. Direction was by Ernest B. Schoedsack and Irving Pichel, with cinematography by Henry Gerrard. It was first shown publicly in November 1932.

Lighting varies from medium mid-key to near low-key with much use of shadows, which appear even in the title sequence as an ornate doorknocker is seen behind the words. A strong side light from the left of the frame casts a shadow on the right side. Typical lighting schemes in the film include strong shadows in the background, some strong key lights, but with reduced fill lights so as to leave some shadow on faces. In the scene in the castle in which the hero Bob Rainsford (Joel McCrae) first meets Eve (Fay Wray) and her brother Martin (Robert Armstrong), all four principal characters (including Count Zaroff, played by Leslie Banks) are usually shown with at least some shading on their faces. Many more striking lighting effects appear. When Bob and Eve discover Zaroff's trophy room with its human heads and are found by Zaroff, the two men argue about Zaroff's 'game'. During this sequence there are two shots of Zaroff with a very dark background and a strong light shining on his face from a very low position, giving the famous 'evil close-up' effect. It is an ECU with the low light throwing dark shadows on his nose and above his eyebrows.

A dark image which does not use high-contrast lighting occurs in the scene in which Bob and Eve are hiding in a cave during the hunt. Several shots show them crouching in the dark, with total shadow in the background and only a very dim light picking out their features. As the hunt progresses, the scenes get darker. The sequence in Fog Hollow gets near to being simply silhouettes moving across an all-grey setting.

Angled shots in *The Most Dangerous Game* are mainly motivated but some unmotivated angles can be noted. In the final fight between Bob and Zaroff there are high-angle shots (which is of course a fairly conventional way of showing a fight). Other high angles include the ELS of Bob and Eve as they cross the main hall of the castle looking for Martin, and the shot of the hounds being released. Low angles include the shots in which the hounds leap over the camera during the chase, and the shot of Zaroff staring towards the couple, after he has avoided their trap. Such angled shots are not frequent, but do occur regularly.

Most of the camera movement in *The Most Dangerous Game* is conventional but one extraordinary movement must be noted. Towards the end of the scene in which Bob meets Eve and Martin, there is a very noticeable unmotivated movement. Eve goes up the stairs. At the top she pauses and looks down towards Bob and Zaroff looking up at her. The camera then makes a quick five-second movement down towards Zaroff, from ELS to MCU. The next shot shows that nobody has moved. The other most notable camera movements in the film occur in the chase scene, to be described below. Just as most of the camera movements are fairly conventional, so are most of the camera distances, although the relatively high number of ECUs is worth noting (for example, during Bob and Zaroff's argument in the trophy room).

As with the other films from this period discussed above, *The Most Dangerous Game* contains some direct-to-camera looks. These are as follows.

1 When first meeting Bob, Eve greets him direct to camera.
2 Zaroff looks direct to camera as he talks to Bob in the trophy room.
3 Zaroff looks direct to camera as he escapes from Bob's first trap.
4 Zaroff looks direct to camera after he thinks he has shot Bob.
5 Immediately after this, Eve and Zaroff exchange direct-to-camera ECUs.
6 At the end of the film, Eve looks back from the escaping boat direct to camera, looking at Zaroff as he tries to shoot an arrow after them.

In addition to these, there is the climax of the chase sequence which is worth noting in detail since it includes a succession of very noticeable moving shots as well as a dramatic lighting effect at the end. This sequence can be taken as a summary of the devices used in this film.

Shot 1 Hounds clambering up a gully.
Shot 2 Travelling two-shot of Eve (MCU) and Bob (MS).
Shot 3 Zaroff (MLS) in the gully.
Shot 4 Zaroff (MLS) in the bushes.
Shot 5 Hounds in the gully (as in shot 1).
Shot 6 Travelling CU of Bob.
Shot 7 Travelling CU of Eve.
Shot 8 Subjective point of view from the position of Bob and Eve, going directly through the undergrowth.
Shot 9 Travelling MCU of Zaroff, direct to camera.
Shot 10 Subjective shot (as in shot 8), but here presumably representing Zaroff's point of view.

Shot 11 Eve and Bob in MCU, looking back direct to camera. They are running forward and the camera is following. It moves into an ECU on Eve, still direct to camera.

Shot 12 Trees and bushes are silhouetted in the foreground, a raging river is in the background. Bob and Eve appear in MLS in silhouette.

Shot 13 Bob and Eve run off in shadow towards a waterfall in the background.

Shot 14 Set-up as in shot 12. A hound jumps at a branch and disappears into the river.

Shot 15 Hound falling into the river.

The complete sequence lasts for just under one minute. The rapid succession of travelling shots (shots 6 to 11) is in itself very striking. Add to this the direct-to-camera shots (shots 9 and 11) and the two remarkable subjective shots (shots 8 and 10) and the result is a highly unusual sequence, rounded off by the dramatic use of silhouettes (shots 12 and 14).

IDEOLOGICAL CONTRADICTIONS IN FILMS OF THE EARLY THIRTIES

These three films are linked by having ideological contradictions at their centre. The contradiction at the centre of *The Public Enemy* is that of the gangster as hero. Tom Powers is a cold-blooded killer and yet he is an attractive figure with energy and personality. The contrast with the film's 'good guy' is instructive. The good character is Tom's elder brother Mike – he studies in his spare time, is honest, volunteers to fight for his country, and attempts to act as Tom's conscience. On the other hand, as played by Donald Cook he is a grim character, taking everything very seriously, and lacking the colour and energy of his brother. On his return from the war, Tom taunts him by pointing out that he has killed people and Mike cannot take this. And of course, even by the end of the film, Tom has retained his mother's love – indeed it is this which makes his death such a powerful climax to the narrative. If there is ambiguity as to whether the central character is a hero or a villain, the style does nothing to resolve this. Its connotations of instability, obscurity and unpredictability allow a great deal of ambiguity. The closing scene is typical. Tom is finally destroyed. He is dead, bound up in bandages and falling to the ground. Yet this is shot from a strong low angle. Unlike a high angle, or a parallel to ground shot (both of which would have tended to emphasise his final indignity) the low angle retains to the last the ambiguity of Tom as a figure to be respected, but also to be feared. In the same way simplicity of interpretation is opposed by the inconsistencies of subjectivity, in which an absent third-person subjectivity is overlapped by the diegetic subjectivity implied by the direct-to-camera looks, and an occasional authorial subjectivity as in the pointed moral of the long mobile shot at the beginning.

The moral ambiguity of the central characters is repeated in *Susan Lenox*, although in this film the questions arise over the status of women and over male attitudes towards them, as the heroine is forced into prostitution by the intransigence of the hero. Helga is trapped from birth and the one point of escape is destroyed by Rodney's lack of faith and understanding. Although the couple end the film together, there is no evidence that Rodney has changed, merely that he is prepared to put up with her. The text thus allows and even invites at least two interpretations – one which sees Susan as a victim surviving the only way left open to her, and another which takes Rodney's side and sees her as a 'fallen woman', too ready to sell herself rather than get 'honest work'.

In this film the ambiguity of the content is reproduced (and thus heightened) by the stylistic signifiers of ambiguity. The style of *Susan Lenox*, as already noted, is not so different from later conventions as is the case in some films of this period. For example the films directed by Josef von Sternberg at this time (some of which – *Blonde Venus*, *Morocco*, *Shanghai Express* – have similarities of content with Susan Lenox) are much more individual. Despite this, it is worth noting how the lighting in *Susan Lenox* works in relation to the content. The contrast between the dark opening scenes and the bright scenes at Rodney's home could be resolved in various ways at the end of the film. Most obviously, an ending in high-key would have tended to confirm the happy ending, linking it to the earlier scenes of happiness. On the other hand, a low-key ending would have tended to emphasise the tragic side of the story and given a stronger resonance to Rodney's warnings about the future. Instead of either of these, the ending is in mid-key, sombre in mood but with the possibility of optimism indicated by Susan's final words – 'I'll make you believe in me'. In other words, the ambiguity is left unresolved at the end, both in terms of the action and of the style.

The ambiguity in *The Most Dangerous Game* is less obvious than in the other films. There is no doubt as to the identities of the hero, heroine and villain. There is, however, a contradiction in the film which is made explicit in the opening scene, before the shipwreck. A discussion takes place as to the relationship between the hunter and the hunted. Rainsford argues that the animals he kills enjoy the hunt, although he tries to evade the question and his answer is immediately brought into doubt by the shipwreck and the sharks which turn the hunters into the prey. This discussion is continued with Zaroff but with the added element of the prize of the hunt being asserted to be the woman. The rest of the narrative works out this discussion with Rainsford and Zaroff both being, at different times, hunter and hunted, and Eve being the prize they both seek. If this seems to be a fairly routine pattern for a horror film and to have few ideological implications beyond the obvious sexism, two points are worth noting. Firstly, the ideological implications may become clearer if the notion of the hunt is replaced with the notion of competition (remembering the common metaphorical uses of the concept of the jungle). Secondly, the uniqueness of this film may be clarified by noting how few horror films explicitly discuss the ambiguities of the hunt. *The Most Dangerous Game* is a horror film which actually discusses the common generic structures of the man/beast and hunter/hunted oppositions, as well as their patriarchal and ethical assumptions. Yet although such issues are raised, they are not resolved. The film ends with one man destroying another, who is actually a beast, and with the hunter, who had become the hunted, becoming the successful hunter again, and gaining the patriarchal prize of the woman. It thus becomes a representation of the contradictions of Rainsford's ideological position.

The relation of this to style can be seen by considering two crucial scenes. The first is the confrontation in the trophy room, which uses low-key lighting, direct-to-camera shots and very close distances. This gives the scene a special emphasis. If the hunter/hunted duality is present in both Rainsford and Zaroff and if the trophy room can be seen as symbolising the darker recesses of Zaroff's mind, then it can function as this for Rainsford as well. In that room he confronts the logical extension of his own role as hunter. The style underlines this situation and also, by the direct-to-camera shots, relates the viewer to this role, momentarily asking him or her to adopt the subject positions of the two hunters. This is developed in the second crucial scene – the climax of the chase (as described earlier) which even more pointedly relates the viewer to the protagonists' viewpoints. In other words, these stylistic devices serve to emphasise the ambiguities of the film, despite the strong closure of the narrative. If the film becomes a discussion of Rainsford's ideological position, then the viewers are deeply implicated in this discussion.

The variety of these three films, particularly as they come from different studios (Warner Brothers, MGM and RKO) and belong to different genres (gangster, contemporary melodrama and horror) should indicate that the contradictions seen in them are not unusual in the films of the period. The lack of, or incomplete resolution of, these contradictions and the way in which stylistic factors emphasise them can be found in many films of 1931–32. The crisis in the nation is thus shown not simply in terms of content (a content that is sometimes displaced to foreign settings) but in narrative structure and cinematographic style.

CONCLUSION

Two final points are worth making. The first is that the analysis of specific scenes and specific shots should not obscure the more general connotations of the style. Perhaps of most importance to the viewer in 1931 was not the interaction of style and content in any one film, but the general appearance of films at the time – the dark tones, the camera angles, the uses of camera movement and the looks direct to the camera. The use of these devices in a large number of films is more significant than their use in any specific films (particularly since this was a period of regular filmgoing, not of occasional attendance at a blockbuster). The second point is that it must be remembered that many films had a far more limited stylistic paradigm. Among well-known films, *Trouble in Paradise*, made at Paramount by Ernst Lubitsch, is an example in which fairly even, bright lighting appears with no surprises in the use of the camera. The point being made in this chapter is simply that a very significant number of films did utilise the broader paradigm and that the availability of this broadened paradigm was due to ideological factors, rather than simply to aesthetic, technological or economic factors. A number of films were able to employ a cinematographic style which expressed the sense of crisis in the early thirties and the ideological contradictions of the period. Very soon, however, that situation began to change.

5 Questioning Subjectivity: *Dr Jekyll and Mr Hyde*

The cinematographic norms of the early thirties, as outlined in the previous chapter, were pushed to near-breaking-point by a few films whose idiosyncrasies indicate the outer bounds of that period's style. Examining such films puts the more conventionally-made films in context, indicating more clearly the stylistic options available. And, as will be argued below, such atypical films were made for specific purposes in specific contexts and as such do not constitute exceptions to the general principle of ideological determination. The film chosen to illustrate this is *Dr Jekyll and Mr Hyde*.

THE MAKING OF *DR JEKYLL AND MR HYDE*

By the middle of 1931 Paramount was in trouble.[1] During the later twenties the studio had greatly expanded its ownership of cinemas, increasing from 300 in 1925 to about 1200 in 1930. However, this position was only achieved at great financial cost and by 1931 Paramount had massive mortgage repayments. The conversion to sound in the summer of 1929 had added to the studio's financial burden. Initially these costs were met by the money made from the first very popular talkies and Paramount even expanded into radio (buying half of CBS in June 1929). However when the Depression finally began to hit Hollywood in 1931, Paramount's position was shown to be extremely vulnerable. It was no accident that the studio which best survived the Depression years was MGM – the one with the least ownership of theatres.

At this point Paramount had not yet adopted the 'producer unit' system of production which was to dominate Hollywood throughout the rest of the decade.[2] Under the producer unit system, a group of associate producers shared the films under production and reported back to the head of production, the executive producer. Instead of this system, in mid-1931 Paramount was still based on a central producer (B. P. Schulberg, working with Jesse Lasky, vice-president in charge of production) and individual directors were often their own producers (especially those with prestige, such as Sternberg, Lubitsch and Mamoulian).[3] Howard Lewis, writing of the situation in 1930, suggested that the Paramount directors 'are free to proceed unmolested with their respective assignments'.[4] Only when the economic crisis hit Hollywood in 1931 did the studio change. In November of that year the producer unit system was adopted with Schulberg in charge of seven associate producers.[5] Schulberg, however, did not take to this system and during 1932 both he and Lasky left Paramount. At the same time as this reorganisation there was a one-third budget cut on all Paramount productions and in all salaries.

Paramount had several very popular stars on its books at this time – Maurice Chevalier, Marlene Dietrich, the Marx Brothers – as well as a number of top-class directors – Ernest Lubitsch, Josef von Sternberg, Leo McCarey, King Vidor and Rouben Mamoulian. Paramount hits of the first part of 1931 included *Dishonored* (d. Sternberg) in March, *City Streets* (d. Mamoulian) in April, *The Smiling Lieutenant* (d. Lubitsch) in May and the Marx Brothers in *Monkey Business* (d. Norman Z. McLeod) in September. Howard Lewis notes that although the box office was suffering at this time, there was still success to be had with the quality film with high production values.[6]

In his autobiographical account of life in Hollywood, B. P. Schulberg's son Budd implies that the principal instigator of *Dr Jekyll and Mr Hyde* was his father. He describes him as thinking that 'here was a chance to please Bob Sherwood [Robert Sherwood, the playwright], make the coveted Ten Best list, and still bring Paramount the profit it needed to keep its head above water in these tricky Depression days'.[7] The director chosen for this production was Rouben Mamoulian.[8] Born in Russian Georgia and educated in Moscow, Mamoulian had arrived in New York in 1923 with theatrical experience already behind him. His Broadway reputation was established with his production of Gershwin's *Porgy and Bess* in 1927 and on the basis of this Lasky invited him to work for Paramount. He made his first film, *Applause*, in Paramount's New York studios in 1929 and then moved west to Hollywood where he made *City Streets* which appeared in April 1931. Both films were innovatory successes, particularly notable for their use of sound. He must have seemed to be the ideal director for a film based on a literary classic which was designed to bring both prestige and box-office success.

Written by Robert Louis Stevenson in 1885 and published in 1886, *The Strange Case of Dr Jekyll and Mr Hyde* had been an instant success, with a stage version appearing as early as 1887.[9] Its obvious visual qualities made it an ideal film subject and S. S. Prawer lists nine silent films based on Stevenson's story.[10] The most famous of these was Paramount's 1920 version directed by John Robertson and starring John Barrymore. Mamoulian's version was in production in the autumn of 1931, thus missing the full force of the November changes at Paramount. The script was prepared by Samuel Hoffenstein and Percy Heath, with art direction by Hans Dreier and cinematography by Karl Struss. It was premiered at the end of December and went into general release in January 1932.

The film was a great success, featuring as one of the top fifteen box-office draws of 1932, and Fredric March won an Academy Award for his performance in the double title-role as well as nominations going to Karl Struss and to the scriptwriters.[11] The film's reputation as a horror film is shown by references to it in Henry

Forman's *Our Movie Made Children* (1933). He describes it as a 'highly intense' movie with 'harrowing visions' and notes that 'a college boy admits that it took him two or three days to get over the fear of dark places inspired by *Dr Jekyll and Mr Hyde*'.[12] Howard Lewis cites it as an example of a film which broke the 1930 production code's rules against showing 'undressing scenes' and scenes of 'brutality and possible gruesomeness'.[13] (This, of course, did not make it particularly unusual – as the effects of the Depression were felt at the box office, many film-makers went for sensational content to bring in the crowds.)

Four points can be made concerning the making of *Dr Jekyll and Mr Hyde*. Firstly, Paramount desperately needed successful movies in late 1931 and were prepared to tolerate a film which bent the established norms of style and content. Secondly, Mamoulian and his team were given a great deal of freedom to develop the film as they wished, partly because of the production system at Paramount and partly because of Mamoulian's own track record. Thirdly, with his established reputation as an innovator, Mamoulian would clearly wish to be seen to be doing something different and original. And fourthly, despite stylistic peculiarities, the film was very successful. Whatever effect the style had, it certainly did not detract from the audience's enjoyment.

STEVENSON'S STORY

The story of *The Strange Case of Dr Jekyll and Mr Hyde* is well-known. Dr Jekyll is an experimentally-minded London doctor who discovers a drug which separates the good and evil parts of the human personality. By drinking it, he becomes a physical and moral monster whom he calls Mr Hyde. Eventually, however, Hyde triumphs over Jekyll as the transformations occur even when the drug has not been taken. The story ends with Jekyll/Hyde's death – suicide in the book, killed by the forces of the law in the film. As far as story content goes, there is only one other major difference between the book and the film. Two female characters are introduced in the latter version – one as Jekyll's fiancée (Muriel, played by Rose Hobart), and the other as Hyde's mistress (Ivy, played by Miriam Hopkins). By 1931 this had become a tradition in cinematic adaptations of the story, two women appearing not only in the famous 1920 version but, as S. S. Prawer notes, as early as the 1912 version.[14] There is, however, another vitally important difference between the film and the book – not in the events of the story, but in the way in which it is told.

The book begins as a third-person narrative but from the point-of-view of Mr Utterson, Jekyll's friend and lawyer. The first eight sections (originally titled but unnumbered) give the story as Utterson discovers it, and include several first-person accounts embedded within it (for example, the eye-witness account by a maid who witnesses Hyde's murder of Carew). These sections give an outsider's view of Jekyll's behaviour and culminate in the discovery of Hyde's dead body.

There then follows a letter written by Dr Lanyon, another of Jekyll's friends, and includes Lanyon's description of the Jekyll/Hyde transformation which he witnesses. Finally Jekyll's own confession is given which fills the remaining gaps in the story.

Right from the beginning this final section is full of indications of the split in Jekyll's personality, with such statements as 'I stood already committed to a profound duplicity of life' and that he was aware of 'this consciousness of the perennial war among my members'.[15] His account of his gradual domination by Hyde becomes a description, however, not of a change of personality but of a loss of identity. He talks of 'this incoherency of my life' – 'I was slowly losing hold of my original and better self.' Hyde becomes another being, addressed by Jekyll in the third person – 'his every act and thought centred on self'. Eventually he even speaks of his better self, Jekyll, in the third person. Hyde 'loathed the despondency into which Jekyll was now fallen'. The destruction of identity is complete in the final sentence in which the 'I' of the narrator is distinguished from Jekyll as much as from Hyde. 'Here, then, as I lay down the pen, and proceed to seal up my confession, I bring the life of that unhappy Henry Jekyll to an end.' Thus the destruction of identity ends in suicide.

The film does not use this method of narrating the story. Events in the film occur in the same order as events in the plot. However, the question of identity is emphasised by the use of the subjective camera. A large number of shots contain looks direct to camera, with the audience being placed in the position of a diegetic character (usually Jekyll or Hyde, but not always). The importance of this will be considered later. Mamoulian himself linked this use of the camera to the importance of the Jekyll/Hyde transformation scenes. 'I wanted to make the transformations a real vicarious experience, more than just a trick. So I used the camera in the first person.'[16]

CAMERA MOVEMENT

Camera movement in *Dr Jekyll and Mr Hyde* is frequent and very noticeable. Few extended scenes lack a complex movement and, in fact, most scenes are structured around such a movement. The statistics of movement over two 100-shot sequences are as follows.

Static shots	47.5%
Reframing movements	18.5%
Pans and tilts	20.0%
Tracking shots	14.0%

Although the term used here is 'tracking shots', most of the mobile shots are actually taken from a dolly, that is, a rubber-wheeled vehicle which is not limited to a set of laid tracks and is thus more manoeuvrable and more free in its use since the tracks do not have to be hidden from the camera's view. Barry Salt notes how the use of the long mobile take from a dolly was common at precisely this period (1931–32) and mentions *Back Street* (1932) with an average shot-length of 23 seconds. *Dr Jekyll and Mr Hyde* has an average shot-length of 12 seconds, slightly above the average for this period (which Salt gives as 11 seconds). Salt also notes a restriction imposed by these dolly shots in that they did not allow the camera to be placed 'much below 3 feet' above the ground.[17]

The complexity of these camera movements can be seen in the eighth shot (not including the opening titles) of the film. This opening sequence is remarkable in that it is all shot from Jekyll's exact point of view and constitutes one of the most sustained uses of the subjective camera in American cinema before *The Lady in the Lake*, the 1946 film directed by Robert Montgomery in which almost the entire film is shot in this way. The eighth shot in *Dr Jekyll and Mr Hyde* follows Jekyll's progress from a room in his house out to his coach. It begins with the camera looking down at the organ music which Jekyll is playing from and then tilts further down to his hands on the keyboard, before looking back at the music again. The camera then pans round to a statuette beside the window and moves towards it. This is followed by a pan round to Poole (Jekyll's servant) standing in the doorway and then a movement forward behind Poole out into the hallway towards the front door. Before reaching this door, Poole disappears through a side door and the camera pans round to a mirror on the wall, showing Jekyll's reflection as he pauses to put on his hat and coat. The camera then pans back round to the front door and advances through it, pausing at the coach as the coachman nods directly to it. As the coach moves forward, so does the camera. The shot ends with a thinly-disguised cut to a view (still from a subjective position) from the coach as it proceeds down the road. The whole shot lasts for one and a half minutes.

This shot is exceptional in that it is more complex than any other shot in the film, but many scenes are built around camera movements. A more typical example is at the end of the ballroom scene at the Carews' house. Muriel and Jekyll have returned to the dance after a love scene in the garden. A shot of the five-piece band dissolves to a shot of the musicians' empty seats, indicating the end of the dance. This begins a sequence-shot lasting for one and a half minutes. From the empty seats the camera slowly pans round to the last guests leaving and then pans back with Muriel and her father as they walk from the door to where Jekyll and Lanyon are standing. After some conversation with all four characters in the frame, the camera moves forward to a two-shot of Jekyll and General Carew as Jekyll tries to get the date of his wedding to Muriel brought forward. This is followed by a pan to a two-shot of Muriel and the General, and then back to a two-shot of Jekyll and the General, who then leaves and the camera moves to a view of Jekyll and Muriel. Jekyll and Lanyon then leave the room and the camera moves into a closer view of Muriel. Finally this dissolves to the next shot of Jekyll and Lanyon walking along a London street (a more conventional parallel tracking shot, but still unusually long at 41 seconds).

It will be clear from these examples that most of the camera movements are very closely tied to the movements of characters. Movements which are independent of characters are very rare. The most memorable is probably the closing shot of the film. After Hyde has been shot dead in his laboratory and has changed back into Jekyll, there is a shot in which the characters are almost static, staring at the body. The camera moves back from a fairly close view of Jekyll framed by his scientific instruments to a position behind a pot boiling over on an open stove (the move is curved, not just a single straight line). The boiling pot has prominently featured as a symbol for Jekyll's repressions several times during the film.

The fact that most camera movement is tied to character movement is a reminder of another aspect of movement – reframing. Small reframing movements following a character as he or she moves slightly do not stand out in this film because they have since become such a standard part of the classical style. Yet, as noted in Chapter 3, in *All Quiet on the Western Front* and in the late silent period reframing was very rare. Its frequent use was an innovation of the early sound period. It might even be suggested that the long camera movements and sequence-shots of *Dr Jekyll and Mr Hyde* show the use of reframing (rather than either a static camera or a very mobile one) was not yet firmly established. By early 1932 most Hollywood films were making frequent use of reframing movements and correspondingly less use of long mobile takes.

CAMERA DISTANCE

The statistics for camera distance over two 100-shot sequences are as follows.

ECU	5%
CU	10%
MCU	16%
MS	20.5%
MLS	26.5%
LS	18%
ELS	4%

These are not particularly unusual figures – the emphasis on MS and MLS is common for Hollywood films of the 1930–50 period. A minor point of interest is the relatively high number of ECUs. In Barry Salt's collection of statistics, the only group of films which consistently use more ECUs than ELSs is the group directed by Josef von Sternberg, one of Mamoulian's stable-mates at Paramount at this time.[18] Having one figure for all ECUs conceals the fact that several of these are eyes-only shots – a closeness of distance which is extremely unusual in Hollywood films during this decade.

In terms of the structuring of camera distances, scenes seldom keep to the convention of establishing long shot leading to conversational medium shots which in turn lead to emphatic close-ups. The scenes with the complex camera movements described above tend to keep at least as far away from characters as MCU, if not MS. There is an obvious practical reason for this. A complex camera movement becomes far more complex if it includes both CUs and LSs due to the problem of depth of focus. An isolated ECU can be shot with a long lens but an ECU in a sequence-shot including LSs requires either extreme deep focus or much careful focus-pulling. Both techniques involve complex lighting set-ups.

Occasionally the structure of scenes is foregrounded. The most obvious example of this is the love scene in the Carews' garden, during their ball. Beginning with the shot in which Jekyll and Muriel sit down on the garden seat, the following sequence occurs. (The camera is static in all shots unless otherwise noted.)

1	LS of Muriel and Jekyll on the seat, beginning with a slight reframing pan	10 secs
2	MLS, same angle	12 secs
3	MS, Muriel and Jekyll in profile	36 secs
4	CU, Muriel and Jekyll in profile	37 secs
5	ECU of Muriel, looking direct to camera	10 secs
6	ECU of Jekyll, looking direct to camera	7 secs
7	ECU eyes only of Muriel, direct to camera	9 secs
8	ECU eyes only of Jekyll, direct to camera	3 secs
9	MCU of the two kissing, camera moves down to a flower, then on to the feet of the butler, then up to a MCU on him	34 secs
10	MLS on Muriel and Jekyll	17 secs
11	LS of the couple standing up	6 secs

Here the effect of moving closer is almost that of the jump cut. There is no change in camera angle, horizontally or vertically, in the first four shots. Only with the cut to the fifth shot is there a change, but here its suddenness (through 90° and followed by several 180° cuts) again foregrounds the changes in distance, creating a very formalised structure (that is, a structure in which the manipulation of form is made very clear to the viewer).

The unpredictability of this kind of structure can be shown by contrasting this scene with others in which the camera distances are used in unusual ways. An example involving the ELS is the scene in which Jekyll returns to his house after getting General Carew to agree to a new wedding date. Jekyll enters his house. The camera is far back, across the large hallway, Jekyll being less than one-fifth of the height of the fame. So far this is not particularly unusual. However, the camera stays at this distance as Jekyll speaks to Poole and moves into a room. Inside the room the camera remains at a long distance – Jekyll now being less than one-third of the frame height. He speaks again to Poole and the distance changes to MLS only after 12 seconds of this shot – an unusually long time to keep an ELS when there is dialogue.

To find a less unusual structuring of distance, the scene following the one just described, when Ivy visits Jekyll, will serve as an example. From the script this would seem to be a conventional scene – two people meet, there is conversation and an emotional climax is reached after which the scene quickly draws to an end. The camera distances are as follows.

1 LS moving to MLS (Ivy at the door).
2 MLS (Jekyll, then Ivy moves into the frame).
3 MS (Jekyll and Ivy).
4 MS (Jekyll seen over Ivy's shoulder as she turns to show him the scars on her back).
5 MCU (Ivy seen over Jekyll's shoulder).
6 MS (Jekyll and Ivy).
7 MCU (Jekyll and Ivy).
8 MLS (Ivy around Jekyll's knees).
9 CU (Ivy).
10 MCU (Jekyll and Ivy).
11 CU (Jekyll seen over Ivy's shoulder).
12 CU (reverse cut from previous shot).
13 MS (Jekyll and Ivy).
14 LS (Ivy).

This is fairly conventional but there are some unusual features, notably the fact that the emotional climax of the scene is in shot 8 which is in MLS, showing the reactions of both characters, rather than the conventional CUs. Despite this, the moves from LS into the medium range (shots 1–8) and then into the closer shots (9–12) with a quick move out again (13–14) are clear and, by this time, have become conventional.

CAMERA ANGLE

High- and low-angle shots are used fairly frequently in *Dr Jekyll and Mr Hyde*. Over the sample of two 100-shot sequences, the statistics are as follows.

High angles	18%
Low angles	27%
Shots parallel to the ground	55%

Of the angled shots, the vast majority are motivated. Only 2 per cent are unmotivated high angles and 5 per cent are unmotivated low angles. The more unusual angled shots are frequently used to frame a character within objects which carry a symbolic charge. Two examples from towards the end of the film should make this technique clear. After Jekyll says goodbye to Muriel for the last time, she collapses on to the piano. The camera cuts to a high-angle shot looking down at her through the three arms of a candlestick-holder on the piano. The angle not only emphasises Muriel's emotional collapse but foregrounds the candles which (with their connotations of religion and bourgeois respectability) now act as dominating and imprisoning bars. The other example occurs in the final scene of the film.

When Jekyll is cornered in his laboratory, a totally unmotivated low-angle shot frames him amongst his laboratory instruments. The police are entering the laboratory at the top of a flight of stairs. Jekyll looks up to them, and the camera (from just below table height) looks up at him. The shot combines the traditional low-angle effect of giving power to the subject (Jekyll is about to say that Hyde, whom the police are chasing, has gone out the other door and so he seems to be in control of the situation) along with a view of Jekyll trapped amongst the scientific apparatus. Such shots, however, form a small minority of the angled set-ups. More typically the angled shots are motivated by character position and make use of the conventional connotations of the high angle giving vulnerability and the low angle giving power.

Perhaps the most striking aspect of *Dr Jekyll and Mr Hyde* is its use of a particular type of horizontal angle – shots in which the subject looks directly at the camera. There are a total of 77 such shots in the film (out of a total of 459 shots, including titles and credits), that is, 16.8 per cent. Despite the fact that such shots are conventionally said to be outside the norms of traditional Hollywood films, discussion of these shots is usually limited to a very few of them, as, for example, in comments by Tom Milne, Peter Lehman, S. S. Prawer and David Bordwell.[19] Of these only Bordwell gets near to recognising their importance in his comment that 'optical subjectivity thus constitutes an important part of the film's intrinsic norm'.[20]

None of these writers, however, indicates the full range of direct-to-camera shots and so it is worthwhile listing them here. An explanation of what is going on will be given later when the ideological structure of the film are discussed.

1. Opening sequence: thirteen shots in which faces look direct to the camera which is in Jekyll's exact position.
2. Jekyll teaching a crippled girl to walk: three direct shots, two of which are from the girl looking at Jekyll, and the other from Jekyll to the girl.
3. Brief MCU of Muriel at the beginning of the Carews' dance.
4. Muriel and Jekyll in the garden: four direct shots, two of Muriel and two of Jekyll.
5. Ivy and Jekyll's first meeting: four direct shots, three from Ivy to Jekyll and one from Ivy and Jekyll towards the door.
6. First transformation scene: ten direct shots consisting of one of Jekyll looking to Poole, one of Jekyll looking in a mirror, seven during the Jekyll/Hyde change, and finally one of Hyde looking in a mirror.
7. Second transformation scene: two direct shots of Jekyll and one of Hyde.
8. Ivy and Hyde in the Music Hall: one direct shot of Ivy looking at Hyde and one of Hyde looking at Ivy.
9. Ivy's rooms: one shot of Hyde looking at Ivy and one of Ivy looking at Hyde.
10. Ivy's visit to Jekyll: two direct shots of Poole looking at Jekyll and one of Ivy looking at Jekyll.
11. Park transformation scene: one direct shot of Hyde.

12 Ivy's rooms: one direct shot of Hyde.
13 Transformation scene with Lanyon: eight direct shots of Jekyll/Hyde changing, with two direct shots of Lanyon and two of Jekyll during the ensuing discussion.
14 The Carews' house: direct shot of Hyde looking at Muriel.
15 Direct shot of Hyde jumping a hedge to escape from the Carews' house.
16 Final scene: eighteen direct shots, of which six are of Lanyon, one of a policeman and twelve of Jekyll/Hyde.

This list not only shows the frequency of direct-to-camera shots but also the number of characters involved. Although the majority of them are shots of Jekyll/Hyde, many other characters look into the camera at some point, including Ivy, Muriel, Lanyon, Poole and a policeman (and, unspecified on the list, at one point in the opening sequence many students in a lecture theatre look directly at the camera). The list also shows that these shots are not reserved for the climactic moments such as the transformation scenes. It is clear that this type of shot is functioning as an important part of the filmic system and not as an occasional special effect. It can then be seen that part of the purpose of the opening scenes of the film is to signal this by an extended and very obvious subjective sequence.

LIGHTING

Lighting in *Dr Jekyll and Mr Hyde* is not as dark nor as highly contrasted as one might expect, given that it is a fantasy horror film. Part of the reason for this can be found in its early date. The darkest of the thirties horror films (such as *The Old Dark House*, *The Most Dangerous Game* and *The Mummy*) came in the period just after this film was issued, that is, in 1932–33. Of the earlier horror films *Dracula* stands out precisely because it is so dark. *Frankenstein*, for example, although certainly a gloomy film, is not as dark as these other later films. Much of the lighting in *Dr Jekyll and Mr Hyde* fits into the classic three-point lighting style with a strong key light, a more diffuse and soft fill light, and a fairly strong back light. This kind of effect is particularly noticeable on the stars, who also get top lighting. Miriam Hopkins (Ivy), in particular, gets the kind of aura of light on the top of her head that is more commonly associated with William Daniels' lighting of Greta Garbo. Occasionally back and top lights are used as key lights. In the garden scene in which Muriel and Jekyll confess their love for each other, the only front light is a weak fill light. Their profiles are strongly lit by two key lights, one on each, from behind, above and on the side, resulting in a mutual glow, as if from within the characters.

Facial shadow is very rare. Its absence may be partly due to the fact that the Jekyll/Hyde division personifies the kind of character split or indecision which half-lighting on the face is often used to signify. In this film the total character changes, not just the facial lighting.

Wall shadows are used rarely but with striking effect. In the final chase, as the police and Lanyon follow Hyde through the streets of London from the Carews' house, Hyde is seen entering the frame in ELS with a small shadow thrown by him on to a building. He runs towards a point just to the side of the camera where there is a strong, low key light. As he moves nearer, his shadow is magnified on to the wall, finally reaching the full height of the two-storey building. This effect is repeated as the pursuers pass by the same building. Another strong use of wall shadow is when Hyde escapes down stairs after killing Ivy. Again the lights are placed low (although not as low as in the previous example) and a strong shadow is thrown on to the staircase wall. An earlier example is when Hyde reappears in Ivy's rooms after she thinks that he has gone forever. In addition to the shadow of Hyde on the wall, there is a harsh light from below which throws Hyde's features into relief. He moves directly towards the camera, passing the light and so moving into darkness, finally blotting out the image altogether. It is worth noting that these three examples (which contain the sharpest shadows in the film) are from scenes without synchronised sound, as was the case in *All Quiet on the Western Front.*

The overall structure of the lighting is linked closely to the Jekyll/Hyde alternation. The opening scenes – Jekyll giving his lecture, his work in the hospital, the dance at the Carews' house – are all fairly brightly and evenly lit. The first dark scene is in the garden when Muriel and Jekyll declare their love. Here the darkness is used in the typically romantic way of isolating the lovers. The next is when Jekyll and Lanyon leave the Carews' house after failing to change the wedding date. This is the scene which introduces the viewer (and Jekyll) to Ivy. Most of the darker scenes are in the second half of the film as Hyde gains the upper hand over Jekyll. The scenes in the laboratory are, in terms of lighting, a mid-point between the lighter scenes in the Carews' house and the darker scenes in the London streets. However, not all Hyde's scenes are dark. Those with Ivy in her rooms are fairly brightly-lit (apart from the example mentioned earlier). This serves to link her with Muriel – both women are linked to Jekyll/Hyde and both are violently affected by this. The comparison is made explicit in several scenes in which a split-screen device is used. A wipe pauses in the middle of the frame, allowing us to see both shots of the transition. Frequently these splits show us Ivy and Muriel. The lighting on Ivy is also bright to keep the 'angel' theme to the fore – Ivy (as mentioned above) is lit with a halo effect and a statuette of an angel is foregrounded when she is murdered.

SIGNIFICATION

The high number of complex camera movements in *Dr Jekyll and Mr Hyde* adds instability to the narrative. It also adds ambiguity, lacking the solidity and certainty of static set-ups. The questions of why the camera is moving and when it will stop are constantly being raised. These movements begin to open up the text, bringing the viewer into the narrative in ambiguous relationships with the central characters.

The camera movements begin to offer an ideological challenge to the neat (perhaps too neat) ending in which the law conquers the deviant scientist. However this challenge is limited to the extent that the movement is, for the most part, linked to character movement. The movement is still excessive (judged by the norms of the time) but is not totally deviant.

Reframing raises a separate issue. It had become a standard part of Hollywood film style by the end of 1931. It clearly makes a big difference to the style but oddly enough few explanations of this change are forthcoming. David Bordwell notes that it was a product of the early sound era. As shooting with multiple cameras developed in 1929–30 to cope with the problems of sound recording, reframing became an essential part of film technique. 'Because the sound was recorded on disks, the action could not be interrupted for a change of camera position, and post-synchronisation was not yet developed.'[21] In order to retain character movement *and* centred framing, reframing movements became essential. But, of course, this does not explain why reframing was retained after the immediate transitional period as sound recording techniques became more sophisticated. Bordwell can only note that 'a device might simply become the preferred paradigmatic alternative, as when reframing replaced quick cutting as a means of keeping moving figures centred'.[22] Earlier in the same book, Kristin Thompson notes that reframing was in use at an earlier period. 'Brief reframing pans became common in the teens; they occurred in about half the extended sample films from 1911, and they remained at about that frequency through the teens.'[23] The 'extended sample' is described earlier as almost two hundred Hollywood films from the period 1915–60.[24] Quite apart from the discrepancy of dates (between 1911 and 1915 as the starting point) it is clear that the number of films from 1911 to 1919 referred to by Thompson is fairly small and of those only half include any reframing movements at all.

The conclusion must be that although reframing was an acceptable and not unusual feature of film style in the period 1911–19, it was certainly not the fundamental feature which it became in the early thirties. Thompson goes on to indicate that by the twenties reframing only 'remained in occasional use'.[25] She offers no reason for this change beyond the presence of planning. In the earlier period lack of planning sometimes led to a character moving out of the centre of the frame and so the camera was reframed. By the twenties such unplanned movements were a thing of the past. Again this does not explain why reframing should later become so much a part of Hollywood film style (and would in fact suggest that the opposite would have been expected).

The lack of adequate explanations for reframing suggests an ideological cause might be at work. Reframing can be seen as the answer to the conflicting needs of narrative excitement and ideological stability. It allows an increase in motion, with the excitement of mobility, while restricting the effects of that instability to a minimum. Film style of the earlier silent period (when very different ideological forces were at work) is not of concern here, but the change in style from the twenties to the thirties can be described as follows. The late silent film uses very few reframes. Then the first sound films need to use reframing to cope with the problems of sound recording, but the technique is retained after the initial need vanishes because it answers a more general ideological need (one which was not so pressing in the pre-sound, pre-Depression years). Reframing is also, of course, bound up with the primacy of the human figure – in particular the figure of the star – and the need to assert this figure's importance over all other aspects of narrative. However, this need could also be answered by quick and frequent cutting, as indeed Bordwell notes.

Camera distance in *Dr Jekyll and Mr Hyde* almost keeps to the established Hollywood style with the emphasis on the MLS to MCU range, keeping characters away from too intimate a relationship with the viewer, but still allowing them to dominate their surroundings. The only disruptive feature (apart from a slight unpredictability as noted above) is the high number of ECUs. This is bound up with the direct-to-camera shots and will be discussed later.

Angles are used in the traditional way, although, as with movement, there is an excess associated with them, there being more and steeper angles in the film than the classical style would usually sanction. The connotations are conventional, dealing with relationships of power, whether moral or social. The shots which contain excess (for example the high-angle shot of Ivy at the piano) draw attention to the device and thus serve an interpellative function, inviting the viewer to adopt a position rather more consciously than usual.

Lighting is used fairly conventionally with the standard connotations of the light/dark opposition. As with the angles, this serves a normalising ideological function – it anchors the style in conventional meanings. The emergence of the classic 'star' lighting system is worthy of particular note. The use of the 'aura' or 'halo' effect serves a variety of functions. Perhaps the crucial one is the construction of the female image *as* image. It is separated from the dramatic context by the clearly non-realistic effect (even within the conventions of thirties realism). The halo effect attracts the gaze, thus implying (a) that the gaze is male and (b) that the object of the gaze – the star – is an object to be worshipped, thus removing the female star from the everyday world of casual efficacy. In this way it can be seen that the star's ideological function of unifying and cancelling contradictions (as argued by Richard Dyer)[26] is present even in the lighting system which unifies the contradictory concepts of the star as object of male desire and the star as pure femininity beyond earthly desire. This also shows how the star image can function as object of attraction for both genders simultaneously. The 'goddess' role offers an image for female viewers which, if adopted as ideal, conceals the reality of the female situation in a patriarchal society.

The structuring of cinematographic elements is very similar to that of *All Quiet on the Western Front*, with a mixture of extreme formalism (as in the garden scene described earlier) and unpredictability (most obviously in the sequence-shots). As in the earlier film, these can work as marks of respectability, as authorial intrusions guaranteeing the artistic quality of the film. The film is thus differentiated from other films, especially other films of the same genre. However, this mixture of structuring devices also, to some extent at least, allows more ambiguity into the text. There is an overall unpredictability about the scenic construction.

SUBJECTIVITY

Dr Jekyll and Mr Hyde shares some of the marks of subjectivity already seen in *All Quiet on the Western Front*, particularly the interpellative devices and the use of occasional shots which break out of the subject positions which the film as a whole constructs. However, the crucial difference is in the use of direct-to-camera shots. These are so important in this film that it would not be going too far to suggest that the filmic system of *Dr Jekyll and Mr Hyde* is in some sense *about* subjectivity.

In the original story, Stevenson undermines the role of the narrator. The novel ends with Jekyll's confession without editorial comment. There is no final summing up from the narrator's point of view. In this way the story appears as a conflict of points of view – a conflict which remains unresolved, even if only in a weak sense. The filmic equivalent of this is to fracture the conventional point-of-view system and this is done largely by the shots in which a character looks directly at the camera.

It would, however, be possible to use direct shots and still present a unified subject position. Robert Montgomery's 1946 experiment *The Lady in the Lake* does precisely this. The almost constant subjective camera always presents the audience with the same character's point of view. By contrast, there is a distinct lack of unity in the point of view presented by *Dr Jekyll and Mr Hyde*. The opening sequence is from Jekyll's point of view, but there is soon a move through the point of view of his students, to that of a crippled girl and then to a more conventional 'third person' point of view. The ending of the film does not neatly tie up the loose ends since the final shot is one of the most obviously unmotivated camera movements in the film – from a point of view which is a third person objective shot but yet which moves in such a way as to question the objectivity and presumed neutrality of such a position (and which also suggests another position – that of the authorial subject overtly manipulating the camera). In this film the point of view of the camera is continually in question. At no time can the point of view of the next shot be safely predicted. The use of ECUs is important here as a highlighting of this process as the whole attention of the viewer is focused directly at the 'mirror' view.

Such a filmic system as this can illustrate the relation between subjectivity, point-of-view and reverse-cut editing. In *Dr Jekyll and Mr Hyde* many of the direct shots are embedded in reverse-cut formations (most obviously in the garden scene). This tends to decrease their disruptive effect. When the overwhelming closeness of the eyes-only ECUs is added there is a very strong effect which, while forcing the viewer into a particular position, at the same time continues to distance him or her from it by its very insistence. The reverse cuts on direct shots (that is, the 180° cuts) in this film bind the text internally but weaken the binding effect on the viewer. Rather than adopting the neutral position of an absent other person, the viewer is pushed into the position of a diegetic character, a position which is accentuated by the ECUs into one of intimacy. The neutrality (and invulnerability) of the 'objective' position is violated. Even in other scenes which do not use ECUs, the 180° reverse cuts on direct shots work towards an exclusion of the viewer. In other words, at certain points in the text, an 'absent person' position is available, while at others the only positions available are those of diegetic characters.

Thus there is disunity in the subjectivity on offer in *Dr Jekyll and Mr Hyde*. The opening assumption of Jekyll's point of view is abandoned but not for one totally coherent position. Even the enfolding force of continuity editing is not always able to overcome this effect. The use of sequence-shots entails a reduction in reverse cuts and although the classic form of subjectivity does not depend only on reverse cuts, it is clearly strengthened by their use.

CONCLUSION

If the technique of manipulation of point-of-view is linked to the narrative content, then an interesting result can be reached. The film shows an ideological conflict. The dominant view, expressed most clearly by Lanyon (of the main characters he is the one who is left holding the stage at the end of the final scene) and satirised in General Carew, is that the sexual freedom represented by Jekyll's experiments (the recognition of repression) is dangerous both to Jekyll himself and to society at large. If the film's point-of-view structure was conventional, then Lanyon's position would emerge without much trouble as the film's viewpoint. Yet this does not happen. Jekyll represents a challenge to this ideological position. He is prepared to follow the path of self-knowledge wherever it may lead, he chafes against society's repressions and he is impatient with the respect for the properties shown by Carew and Lanyon. He unashamedly talks of marriage as an excuse for being permanently 'drunk with love' and sees conventions as bonds rather than bounds.

Although such views are clearly deviant in the context of the film's version of Victorian England, they are not necessarily views which Hollywood in the 1930s would reject – they can quite easily be romanticised. However, their rejection by the dominant ideology of the film is made clear by the fact that Jekyll's experiments lead not just to unrepressed sex, but to totally asocial physical violence. And, of course, Hyde is revealed to be very animal-like.

Thus the film's principal ideological project works to associate sexual freedom with physical violence, seeing the need to repress both for the sake of society. In addition to this, it is important to note the representations of women, class and professionalism which the film endorses, all of which are firmly within the dominant ideology of the times. The sub-codes of signification and the structuring of the cinematographic elements do little to hinder this ideological position. However, the system of the subjectivity (or rather, subjectivities) allows for a variety of subject positions at different points in the film and thus tends to subvert, or at least loosen, the force of the ideological project. What is found, then, is a Hollywood film which through its cinematographic style is able to open up cracks in the ideology of the content. As such it is clearly an exceptional film, but the facts of its critical and box-office successes should not be forgotten. Exceptional it might have been, but it was an exception which the audience of Depression America responded to enthusiastically.

6 The New Deal in Hollywood, 1933–35

The key year in the decade for the United States was 1933, the beginning of the New Deal. But if the mid-period of the decade began there, its ending is not so clear and, as far as Hollywood is concerned, the middle of the decade can be divided into two periods – a transitional period from 1933 to 1935, which saw a move away from the style and content of the early thirties, and the period from 1936 to 1938 which saw the classic restrained style at its peak. By 1939 important changes were once more taking place. But the background to all of this was the state of the nation.

THE ECONOMY AND SOCIETY IN THE MID-THIRTIES

As before, the three indicators of unemployment, gross national product and economic growth will show the general trends in the United States' economy in the period from 1933 to 1938.[1]

Year	*Unemployed*		*GNP*	*Economic growth*
	Number	*% Workforce*		
1933	12 830 000	25.2	55.6*	−4.0**
1934	11 340 000	22.0	65.1	−2.5
1935	10 610 000	20.3	72.2	−1.2
1936	9 030 000	17.0	82.5	0.2
1937	7 700 000	14.3	90.4	0.6
1938	10 390 000	19.1	84.7	0.1

*Gross national product is expressed in $billions by 1975 prices.
**Economic growth is calculated from a 1926 base.

These figures show the gradual climb out of the bottom of the Depression in 1933, and also the later recession which hit the economy in autumn 1937, with recovery coming the following summer. The cinema statistics are as follows.

Year	*Average weekly film attendance* (millions)	*Box-office receipts* (millions of dollars)
1933	60	482
1934	70	518
1935	80	556
1936	88	626
1937	88	676
1938	85	663

As noted in Chapter 4, these are not necessarily reliable figures, the attendances coming from *The Film Daily* and the receipts from the Motion Picture Association of America. However even if, as seems likely, these figures are exaggerated, they at least indicate the general trend which is the important point in the present context, with a steady improvement from 1933 until the recession of 1937–38.

These statistics can give an inaccurately optimistic view of the mid-thirties. The economy was certainly improving but only from a very low point at the beginning of the decade and it was not until the war that it fully recovered. To make matters worse, 1934 saw the beginning of the 'dust bowl' centred around Oklahoma and Arkansas, created by the combination of drought and poor farming methods. Over the next few years families travelling west from the dust bowl states to the Pacific coast became a common sight. The work of photographers such as Walker Evans, Ben Shahn and Dorothea Lange for the Farm Security Administration shows graphically how extremes of poverty and hardship still existed in 1938. M. A. Jones has noted how the New Deal had not, by the end of 1934, 'brought much benefit to some of the most disadvantaged groups: sharecroppers, small farmers, the rural unemployed, the old'.[2]

Roosevelt's New Deal administration is often seen as having two phases.[3] In 1933 and 1934 it was mainly concerned with recovery and a large number of acts were passed such as the Federal Emergency Relief Act (authorising the payments of federal grants to states for unemployment relief), the Agricultural Adjustment Act (aimed at controlling farm production), and, most famously, the National Industrial Recovery Act (a wide-ranging attempt to revitalise industry while protecting workers), all of which were aimed at stemming the Depression. In addition, various bodies were set up – the Civilian Conservation Corps, the Public Works Administration and the Tennessee Valley Authority – which represented direct federal intervention in the economy.

The second phase began in 1935 and is usually seen as being more concerned with reform than with recovery (although the distinction should not be exaggerated). The first year of this phase saw a Social Security Act establishing a national pension scheme and unemployment insurance, a Wealth Tax Act, and the National Labor Relations Act (usually known as the Wagner Act) which protected trade union rights (Jones has described this as 'perhaps the most sweeping reform of the New Deal period').[4] 1935, however, was also the year in which the Supreme Court began to attack the New Deal. In May it declared the National Industrial Recovery Act to be unconstitutional. The Agricultural Adjustment Act went the same way in January 1936.

Roosevelt was, however, returned in the 1936 presidential election with an even bigger percentage of the popular vote but his second term was marred by the recession of 1937–38, caused, according to Jones, by Roosevelt's attempt to balance the budget by cutting federal spending.[5] Some important reforms were pushed through, such as a revised Agricultural Adjustment Act and a Fair Labor Standards Act (which established minimum wages and maximum hours), both passed in 1938. Generally, however, the New Deal was being seen by many as an exhausted idea. Jones notes how Roosevelt's annual message to Congress in 1939 marked the change: 'For the first time since coming to office he proposed no new reforms, emphasizing instead the threat to world peace and the need for national defense.'[6]

New Deal policies have been seen by some as revolutionary and socialist, and also (given Roosevelt's election victories) very popular. To balance this, two points are worth remembering. The first is that the New Deal essentially shored up the American economy, rather than dramatically changing it. Indeed some historians have argued that Roosevelt's ultimate achievement was to protect American capitalism from radical change.[7] The contrast with the situation in Britain after 1945 is instructive. There were no American equivalents to the National Health Service or the nationalised industries.

The second point to remember is that the New Deal had many enemies. Two quotations from Studs Terkel's *Hard Times* will indicate the strength of the feelings of animosity which were aroused. The first speaker is Charles Stewart Mott who, at the time of the Terkel interview in the late 1960s, was on the board of General Motors. 'Someone said to me: Did you see the picture on those new dimes? It's our new destroyer. It was a picture of Roosevelt. He was the great destroyer. He was the beginner of our downhill slide. Boy, what he did to this country. I don't think we'll ever get over it. Terrible.'[8] Jerome Zerbe was a New York society photographer in the thirties. Terkel asked him what the New Deal meant to him. 'It meant absolutely nothing except higher taxation. And that he [Roosevelt] did. He obviously didn't help the poverty situation in the country, although, I suppose . . . I don't know – New Deal! God! Look at the crap he brought into our country, Jesus!'[9]

Another feature of this period was the popularity of demagogues outside of the main party structures – Huey Long in Louisiana, Charles Coughlin on the radio and Francis Townsend appealing to the aged. Long in particular had to be taken seriously by mainstream politicians. At the time of his assassination in September 1935, he had been intending to run for the presidency.[10]

What all this indicates is that the New Deal period was one of a rather problematic reassertion. The immediate crisis over, the need was to re-establish stability – ideological as well as economic. The improving economy gave the ideological superstructure the impetus it required to regain the confidence that had been lacking earlier in the decade. However, the crisis had been so severe that a return to pre-Depression ideological certainties was not possible. New Deal rhetoric represents an attempt to redefine the central ideology of American capitalism – an attempt that was only partially successful as the above quotes from Mott and Zerbe show. Only when the United States entered the war after Pearl Harbor was bombed in December 1941 did the country regain some kind of ideological unity.

HOLLYWOOD AND THE PRODUCTION CODE

This situation put a certain pressure on Hollywood in contrast to the freer (that is, more ideologically autonomous) world of literature which gave a rather different view of the conflicts of the period. John Dos Passos completed his trilogy *USA* in 1936, and John Steinbeck published *Of Mice and Men* in 1937 and *The Grapes of Wrath* in 1939. Drama included Clifford Odets' *Waiting for Lefty* (1935), Lillian Hellman's *The Children's Hour* (1934) and Thornton Wilder's *Our Town* (1938), quite apart from Orson Welles' black production of *Macbeth* in 1936. Hollywood, tied more closely to the economic base (largely because of the scale of the finance needed in film-making and the consequent need to attract a large audience), did not reflect the conflicts in the same way. To understand what happened, the most useful approach is to look at the circumstances surrounding the strengthening of the Production Code in 1934.

The movement towards greater censorship in Hollywood is linked in an important way to the Payne Fund studies, most of which were published during 1933.[11] These were empirical studies done by the Motion Picture Research Council under W.W. Charters, with finance from the Payne Fund. The research was carried out between 1929 and 1933 and, as the titles will indicate, were mostly concerned with the effects of films on children. The eight titles published during 1933 were *Movies and Conduct, Movies, Delinquency and Crime, Motion Pictures and Youth, Getting Ideas from the Movies, Motion Pictures and Standards of Morality, Motion Pictures and the Social Attitudes of Children, Children's Sleep* and *The Social Conduct and Attitudes of Movie Fans.* Three more volumes – *Children's Attendance at Motion Pictures, The Emotional Responses of Children to the Motion Picture Situation* and *The Content of Motion Pictures* – were published in 1935. In addition to these, Henry James Forman published *Our Movie Made Children* in 1933 as 'a popular summary'.[12] Jowett notes that it 'caused a mild sensation, was reprinted at least four times and was widely reviewed and discussed'.[13] The direction of Forman's argument will be clear from the following quotation. 'If, therefore, as appears, the movies act as a system of education for large portions of the population, then we must not delay in taking the necessary measures to treat them as a system of education.'[14] In other words, the Payne Fund studies were taken as evidence that greater censorship was needed, particularly where children were concerned.

Forman's argument did not go unanswered. Raymond Moley, a Roosevelt advisor, published *Are We Movie Made*? in 1938. This book, written 'at the suggestion of representatives of the motion picture industry', is heavily based on the philosopher Mortimer Adler's *Art and Prudence*, published the previous year.[15] As well as a damning criticism of the research in general and of Forman's book in particular, Moley's work is a defence of the film industry's ability to control itself. His main suggestion is for 'the discrimination among motion pictures of those which can be exhibited to any audience and those which are for adults only'.[16] The significance of this argument between the Payne Fund researchers on the one hand, and Adler, Moley and the Hollywood studios on the other, is that it shows both the seriousness with which these criticisms were treated, and how the terms of the debate were set. Moley's final quotation from Adler is as follows. 'The agitation about the movies, resulting in extensive scientific research, has not resulted in any findings that are nearly as much a cause for alarm as the terrible confession of inadequacy which the agitation itself so plainly bespeaks.'[17] The 'confession of inadequacy' relates very closely to the inadequacies of traditional ideological positions in the early thirties, specifically their inability to account effectively for the Depression and the resulting social upheaval.

A year before Moley's book, Martin Quigley (one of the authors of the 1930 version of the Production Code which the 1934 agreement enforced) published *Decency in Motion Pictures*, taking a similar line to Moley, although being an even more explicit defender of the industry and the Production Code. Quigley makes the typically ideological move of asserting his standards as those of the vast majority and as the only natural ones. 'There are and must be rules to every game and these are the rules in the game of life which since the dawn of civilization – and still – have the acceptance of an overwhelming proportion of mankind.'[18] His conclusion makes quite explicit his aim of 'the maintenance of public and private morality'.[19] Moley himself was later to describe this as Quigley's 'splendid little book', seeing it as giving the philosophical basis of the Production Code.[20]

All of this makes clear the discussion that was going on at the time, and which can now be seen as the reassertion of a traditional ideology after the chaos of the early thirties. The arguments of the time serve to limit the area of debate (the only positions seem to be that of those who wanted government legislation and that of those who wanted industry self-regulation) and to assume a shared and unproblematical ethical code. This argument soon moved beyond published polemics. In November 1933 Joseph Breen was put in charge of administering the Production Code at the Motion Picture Producers and Distributors of America (usually known as the Hays Office) and in the following January he became chairman of the studio-relations committee. This became the Production Code Administration in July 1934. In April 1934 a committee of American Roman

Catholic bishops started a 'League of Decency' and within a few months, according to Moley in his official history of the Hays Office, it had several million supporters, with praise from Protestant and Jewish leaders. In June Breen managed to get studio agreement for a 'Resolution for Uniform Interpretation'. This enforced adherence to the Code, asking members not to distribute, release or exhibit any films which had not had MPPDA approval.[21]

It is easy to over-estimate the effects of this agreement and so three points are worth noting. Firstly, the MPPDA forced changes in films before 1934. Richard Maltby's account of the production of *Baby Face* deals with events between November 1932 and May 1933, and shows changes being made to the script on MPPDA advice.[22] Secondly, by 1933 film content was already retreating from the sensationalism evident in 1932. Thirdly, even after the summer of 1934, code enforcement was not absolute. Nick Roddick writes (in a discussion of Warner Brothers' *Anthony Adverse,* released in late summer 1936) 'however absolute they may sound, Breen's statements are in fact the basis for negotiations'.[23] The MPPDA was, essentially, offering advice on how to make films acceptable to a wide public and therefore to avoid censure from any significant group. This is not intended as a denial of the importance of the Code's enforcement in 1934. There is no doubt that it was a significant event. However, rather than being seen as the essential cause of change, it should be regarded as a symptom or even a landmark of a broader change which began before the summer of 1934.

TECHNOLOGICAL DEVELOPMENTS IN THE MID-THIRTIES

This is a suitable point at which to return to the question of cinematographic style and technology. The most important development in lighting units and during the decade was the Mole-Richardson spotlight with Fresnel lenses, available in 1935.[24] However, the increased control which this lighting unit gave to the cinematographer could be used in two ways – to create more complex lighting effects, or to create the same effects as before but more cheaply. As Bordwell notes, 'after 1935 it was not uncommon for cameramen to mix incandescents with the improved Mole-Richardson arcs, again in the name of control, economy, and efficiency'.[25] Thus high-key lighting, in the later thirties style, was developed *against* financial pressures. Not only was mid-key lighting quicker to set up and cheaper to run (with electricity as a major cost for any studio film) but it meant that less care and detail had to go into the sets.

The other factor which affected the light image was the film stock. The one constantly changing element of cinematographic technology during the thirties was the increasing speed of film stocks.[26] In 1930 the standards were set by Eastman Type I and Type II Panchromatic negative, both first introduced in 1928, and with speeds of around 20–25 ASA.[27] The main developments during the rest of the decade were as follows.

1931 Eastman Super Sensitive Panchromatic Negative (created, according to Bordwell, for low contrast Mazda lighting).[28]
1934 Agfa-Ansco Super Panchromatic negative introduced into the USA, with a speed of 32 ASA.
1935 Eastman Super X Panchromatic Negative, with a speed of 40 ASA.
1938 Agfa Supreme, speed: 64 ASA.
Agfa Ultrapan, speed: 120 ASA.
Eastman Plus X, speed: 80 ASA.
Eastman Super XX, speed: 160 ASA.

Of these, the most important were Eastman's Super X in 1935 which improved picture quality but was a low-contrast stock[29] and Eastman's Plus X in 1938 which greatly improved definition and became the Hollywood standard for some years. Ogle notes that the higher contrast of Plus X was such that many cinematographers 'were to experience real difficulty in lighting sets properly for the new film'.[30] These changes have usually been discussed in the context of deep focus (as in the above quotes from Ogle and Bordwell) but it should be clear that they are relevant to the more general issue of the development of lighting styles.

It may seem as though the introduction of Super X in 1935 provides the answer to the time lag (noted in Chapter 4) between developments in lighting technology in the early thirties and the development of the high-key style in the mid-thirties. Two points can be made about this. Firstly, simply to note the dates of technological developments does not explain why those particular developments happened at that particular time. Secondly, there is no necessary link between improved lighting, faster stock and high-key effects. These developments could just as easily have led to a low-key style (after all, discussions of the origins of deep focus are usually pointed towards *Citizen Kane* or the films of William Wyler). In other words, simply noting historical events does not provide an explanation of those events.

One other major development at this time must be mentioned – the development of colour film. The Technicolor Corporation first marketed a three-strip colour camera in 1932.[31] Instead of the earlier two-strip process, using only combinations of red and green, the three-strip process used red, green and blue, thus being able to produce a much more life-like range of tones. It was first seen by the public in a Walt Disney cartoon short in 1932 (*Flowers and Trees*) but its use for non-animated films was delayed until 1934. In that year it was used for a short film (*La Cucaracha*) and for sequences in four feature films, but its first use for a complete feature was in 1935 (*Becky Sharp*). This was followed by a gradual expansion during the rest of the decade. In 1936 there were four features completed using the technique. In 1937 there were five, with another two being made in Britain and sequences in two others. By 1938 the number of complete features had risen to eleven (one of which was animated – *Snow White and the Seven Dwarfs*), with three in Britain and sequences in one other. There was a slight fall to eight in 1939, plus two in Britain and sequences in four others (including *The Wizard of Oz*).[32]

Although clearly an important development in the overall history of cinema, the few films made in colour in the thirties make it a relatively unimportant aspect of the decade's stylistic options.

What is found during these years of change is the gradual emergence of a cinematographic style based on high-key lighting, reframing, strongly motivated camera movement and angle, and very predictable structuring, particularly of camera distances. This restrained style (restrained in its deliberate use of a narrow range of stylistic options) had fully emerged by 1936 and, as argued above, although the developing technology had a role to play in this change, it was not in itself a sufficient cause for that particular style. Only when the style is related to the ideological conditions of the period is a sufficient reason found. As ideological reassertion takes place in the country at large after the trauma of the pre-New Deal years, both the content and the style of Hollywood films become more restricted, partly due to the emergence of pressure groups such as the League of Decency but also, at least as far as cinematographic style is concerned, due to a less consciously felt need for stability, clarity and simplicity. The steps towards the emergence of this style can be seen during the years 1933–35 and can be traced by considering four films – *42nd Street*, *Viva Villa!*, *Bride of Frankenstein* and *Les Miserables* – which show the various conflicts and changes which occurred in the transition from the style of the early thirties to the style of the middle of the decade after 1935.

42ND STREET

Although *42nd Street* was released by Warner Brothers in March 1933 (to coincide with Roosevelt's inauguration and make the most of the advertising slogan 'Inaugurating a New Deal in Entertainment') it was actually completed by early November 1932 and previewed in January.[33] It was directed by Lloyd Bacon, with Busby Berkeley in charge of the musical numbers. The cinematographer was Sol Polito. It was nominated for an Academy Award (Best Picture) and reached the top eleven box-office hits of 1933 (as did its follow-up of June, *Gold Diggers of 1933*).[34]

Lighting in *42nd Street* is unremarkable. It is in mid-key and the very little shadow in the film is highly motivated. This, along with the fairly predictable structuring of scenes, can make *42nd Street* seem a typical mid-thirties film. However, there are some interesting deviations from this restrained style. The film includes a number of extreme high-angle shots (quite apart from the ones in the musical numbers). Some of these are backstage shots, looking down from the wings. Such set-ups, sometimes going beyond the backstage view, became a standard feature of the Warners' backstage musicals, and can be seen in such films as *Gold Diggers of 1933* and *Footlight Parade*, the third in the sequence, released in October 1933. These extreme high angles are usually unmotivated and, although brief, tend to disrupt the established relationship between the viewer and the film.

The other points of cinematographic interest in *42nd Street* occur in the musical numbers and in the opening montage sequence. This opening sequence is as follows.

Shot 1	Travelling aerial view of Manhattan.
Shots 2–9	Street signs on 42nd Street. Five of these shots are low angles and one is a high angle.
Shot 10	Low angle of a man at a telephone.
Shot 11	Extreme low angle of a girl on a telephone answering the man in the previous shot.
Shot 12	Slight low angle of a man at a telephone.
Shot 13	A couple seated together speaking to each other.
Shot 14	ECU of a mouth, with direct-to-camera MCUs superimposed in miniature.
Shot 15	Extreme low angle of telephone engineers under a manhole.
Shot 16	MCU of a telephonist.

This sequence lasts for 39 seconds. The dialogue in shots 10–16 is simply the repetition of 'Jones and Barry are doing a show'. The high and low angles of this sequence and the direct looks in shot 14 are not particularly unusual in the context of a montage sequence, although this sequence is not the rapidly superimposed series of shots normally thought of as Hollywood montage (as in the style of Slavko Vorkapich's work at MGM). The force of this is not great since it is fairly short but its position at the beginning of the film tends to delay the creation of a unified subject position until the following scene.

Of the three musical numbers which form the final part of the film, the first ('Shuffle Off to Buffalo') is unremarkable. It includes girls looking direct to camera but this is firmly within the conventions of the staged musical performance. The second number ('Young and Healthy') is what is now thought of as a typical Busby Berkeley creation, with the camera travelling through the open legs of chorus girls and overhead shots of dancers formed into circular patterns.

It is worthwhile emphasising just how unusual these overhead shots are. They are now so well-known that they arouse little surprise, yet they conflict with the subject position which the rest of the film has been constructing. The overhead shots transform people into patterns, thus breaking the viewer's involvement with the characters. This feature is emphasised by the fact that these numbers are set on a diegetic stage in front of a diegetic audience, though the patterns cannot be seen by that audience. Despite this move away from the conventional construction of subjectivity, these numbers have remarkably little disruptive force. This is probably due to their explicit appeal to male voyeurism. It suggests a subject position which is even more regressively patriarchal than that of the rest of the film. Thus, although the subjectivity is different, it is related, being, as it were, a position hidden behind the subjectivity of the rest of the film and occasionally emerging elsewhere (or at least being hinted at), for example when the camera examines the girls' legs at the audition or when Abner Dillon (Guy Kibbee) is seen ogling at them. To put it another way, the fantasy evoked by the musical numbers is implicit in the film as a whole, with its emphasis on males controlling female performances. Thus although the film constructs more than one subject position, these positions are linked in an important way, and this tends to reduce any disruptive effect which this conflict might have.

The final musical performance of the film (the title number) is remarkable mainly for the number of mobile crane shots which it contains – 14 out of 26 shots in a five-and-three-quarter minute sequence. Travelling shots occur elsewhere in the film but most are with character movement. Some are quite striking, such as the 10-second crane shot towards the stage at the start of the audition scene or the 35-second back-and-forth movement following Marsh (Warren Baxter), the show's director, as he gives the cast their introductory pep talk. However, the movements in the final number outdo any of these earlier shots. They add a dynamism to the performance which matches the lyrics, describing 42nd Street as full of action and rhythm. This match between camera movement and the content of the number tends to limit the force of the movement by giving some kind of motivation to it. The instability which the movement signifies is explained as simply the instability of life on 42nd Street.

It will be seen from these comments that although *42nd Street* is well on the way to the more restrained style of the middle of the decade, it still contains some features which link it to the earlier style. In the next film to be discussed a different stylistic mix is present and one which gives a stronger impression of conflict.

VIVA VILLA!

Viva Villa! was released by MGM in April 1934. The credited director is Jack Conway but Howard Hawks began the production. Apparently Conway took over when Hawks was recalled from Mexico by Louis B. Mayer to act as a witness against actor Lee Tracy.[35] The cinematographer was James Wong Howe who, according to Rainsberger, was responsible for the film's visual style.[36] Howe was assisted by Charles Clarke as second-unit cinematographer. Howe claimed to have shot 90 per cent of the film, although Rainsberger (generally a witness in favour of Howe) regards this as exaggerated.[37] *Viva Villa!* received one Academy Award (for John Waters, the assistant director) as well as two nominations (for Best Picture and for Ben Hecht's script).[38]

The cinematography of *Viva Villa!* is very striking but this is due almost entirely to the lighting. There are not many strongly-angled shots and the ones there are, are usually strongly motivated. One example is in the final scene in which Pancho Villa (Wallace Beery) is shot. It includes a very high-angle shot of him in the street, seen from the murderer's point of view. The camera looks down from a high window. Not only is the angle strongly motivated by character position but the silhouettes of the murderers are seen in the foreground leaving no doubt as to the source of the point-of-view. This is typical of the high-angle shots in the film. The only unmotivated high angles are conventional views of a large area. Low angles are also usually motivated by character position, for example, looking up at somebody on horseback. Camera movement is mostly with character movement or is a mobile establishing shot. Thus at the beginning of the trial scene near the start of the film, the camera moves along a line of victims and up to the trial judge, establishing the spatial relationships in the scene. The scene with the most movement is the final battle scene which includes the parallel tracking shots familiar in war films (as was seen in *All Quiet on the Western Front*). However, even these shots are all fairly brief. Camera distance is kept further than ECU except in the montage sequences. The 'Villa wants you!' montage showing the re-gathering of Villa's army after the murder of Madero, includes a couple of ECUs, but only as a very short part of a longer sequence. The structuring of scenes is fairly conventional and there are no direct-to-camera looks.

The style of *Viva Villa!* is thus fairly restrained, except for the lighting, as mentioned earlier. The lighting scheme is typified by an attempt at naturalistic source lighting (Rainsberger notes that Howe consciously aimed for this whenever he could).[39] In exteriors this results in some strongly shadowed shots, such as in the opening scenes. When Pancho's father is being whipped to his death, not only are there strong shadows cast on the ground but the overseer of the punishment and the watching peasants all have shadows on their faces, cast by their hats. This is followed by a night scene in which Pancho kills his father's murderer. Strong shadows dominate the set. Victim and avenger are both shown either in silhouette or with an edge of lighting. Even the interiors are occasionally lit in a very striking way. The first view of Rosita (Katherine de Mille), prepared in typical fashion by a wide-eyed stare from Pancho, is a remarkable MS. Her body from waist to neck is strongly lit, but her face is in light shadow with an edge of light down either side. The background ranges from light shade to very dark shade. Although not 'realistic' lighting in any obvious sense, the play of light and shade creates a version of interior lighting less unnatural than if brighter high-key lighting had been used. Another scene of particular note is that in which Pancho whips Teresa (Fay Wray). Throughout this scene high-contrast lighting on Pancho is contrasted with more even lighting on Teresa (although occasional shadows are seen on her, especially after Pancho has knocked her to the ground). The whipping itself is seen as shadows on a wall. In many other scenes low-key lighting is found, sometimes in the form of foreground shadows.

These examples should be enough to show that the use of lighting in *Viva Villa!* is far more complex than the use of the other elements of cinematography. These other elements belong fairly firmly within the restrained style of the mid-thirties, whereas the lighting is unusual enough to add a degree of ambiguity. This is particularly interesting in *Viva Villa!* since the film shows so much confusion in content. *Viva Villa!* both supports revolution and condemns it. It asserts that the

Mexican peasant are capable of making a revolution and yet it represents their leaders as, at best, infantile. The film overtly condemns violence, yet humour is made out of Pancho's battles. Wallace Beery plays Pancho as a simple, childlike figure, yet the film includes, in the whipping of Teresa, one of the most overtly sadistic scenes in any Hollywood film of the thirties. The low-key lighting, in its ambiguity and obscurity, tends to emphasise these darker, contradictory elements, resulting in a profoundly ambivalent film.

The reasons for this (apart from Hollywood's continual problem of how to portray a 'good' revolution) may be connected with the conditions of production. Shooting in Mexico, far from the studio, and having a change of director may have allowed Howe, always a strong-willed character, the space to establish his own style. In 1934 he was one of the most innovative cinematographers in Hollywood, having just made *Transatlantic* and *The Power and the Glory* and, later that year, shooting *The Thin Man*. Having been able to light the film according to his idea of realism (and thus working to his own professional ideology), Howe may have unintentionally underlined the ideological tensions implicit in it.

BRIDE OF FRANKENSTEIN

Bride of Frankenstein was not quite the end of Universal's first cycle of horror films but, appearing in May 1935, it was something of a final flourish, particularly for director James Whale (who had also directed the original *Frankenstein* in 1931). The cinematographer was John Mescall. *Bride of Frankenstein* uses many of the cinematic devices which have already been noted as typical of the early thirties – low-key lighting, unmotivated angles, extreme distances, and fluid camera movements. In addition, it makes much use of canted camera angles in the final scenes. Some examples will make this style clear.

The first appearance of Dr Pretorius (Ernest Thesiger) is immediately preceded by Elizabeth, Frankenstein's wife (Valerie Hobson), telling of her hallucinations of a figure of death. Following this a dark, cloaked figure is seen hammering at the door, framed between a strong shadow on the left and the shadow of a tree on a wall to the right. The camera is at a slight low angle. There is then an ELS of the cavernous hallway with Minnie, the housekeeper (Una O'Connor), appearing in the distance. Foreground silhouettes and an outsize wall shadow behind Minnie fill the frame. She moves across the hall and the camera tracks with her. A brief MLS of Minnie is then followed by a continuation of the previous shot and a MS of Pretorius outside. The next shot shows the door opening, viewed from inside, with Pretorius slowly revealed in MCU. Half his face is completely in shadow and the other half is partially shaded. As the door moves further back the shadow is removed and his face is revealed, harshly lit by slightly low-placed lights. A reverse-cut MS of Minnie is then followed by the previous set-up. This time Pretorius moves forward (at the words 'a matter of *grave* importance') into what becomes a low-angle ECU with low-key lighting.

Another example is the scene in which the monster (Boris Karloff) stumbles through a graveyard. It is lit in low-key with foreground shadow, low angles and several tracking shots. The movements and angles are as follows.

Shot 1 29-second parallel tracking shot accompanying the monster through the graveyard.
Shot 2 Slightly canted low-angle shot of the monster toppling a gravestone.
Shot 3 The monster descends into a vault.
Shot 4 The villagers in the graveyard, seen through a six-second parallel tracking shot.
Shot 5 Very low angle looking up at the villagers as they pass the vault.
Shot 6 Continuation of shot 4, with the camera tracking with the villagers through the graveyard, at a slight high angle.

In all of these shots faces appear in shadow and the foreground has near-silhouettes of gravestones and/or withered trees.

Despite these examples, the film as a whole does not have the oppressive mood of many early thirties films which use the same devices. This is most obvious when it is compared with the 1931 *Frankenstein*. There are four reasons which might be suggested for this. Firstly, the predominant style is less extreme than these examples might suggest. Much of the lighting is mid-key and although there are many high- and low-angle shots, most of them are at least weakly motivated (for example, low-angle shots of the monster who is taller than anybody else). There are many scenes containing the stylistic excesses described above but they still form a minority of the whole film. Secondly, the bulk of the film is overtly presented as narration. At the beginning Mary Shelley (played by Elsa Lanchester, who also plays the monster's mate) is seen talking with Byron and her husband. Byron reminds them of the original story, as 24 shots from the 1931 film are shown in a 42-second sequence. Then Mary starts to tell them of the continuation. This places a frame around the story, presenting it as a character's fantasy and thereby motivating not only the excesses of the plot, but also the stylistic excesses. The third reason for the less oppressive impact of *Bride of Frankenstein* is that it is, at least to some degree, a comedy. Minnie in particular is played as a comic character, as are Dr Pretorius and the Burgomaster (E. E. Clive).

The final reason (which is related to the previous one) is the self-consciousness of the film. There is little attempt at even a studio realism. When the monster is captured, it is in a forest of trees that have no branches and little undergrowth. The high-point of this stylistic mannerism is the final part of the film when the monster's mate is created. A series of canted shots (one 23-shot sequence includes 18 canted shots, one of which is a 13-second shot with a pan) leads up to the female monster's awakening, marked by an eyes-only ECU. The structuring of the following scene is very unpredictable.

Shot 1 LS of Pretorius, Frankenstein (Colin Clive) and the female monster.
Shot 2 MLS of the female monster.

Shot 3 MCU of the female monster.
Shot 4 Low-angle ECU of the female monster.
Shot 5 Low-angle MLS of Pretorius.
Shot 6 Set-up as in shot 1.

There is no character movement in these shots, other than the monster's mate moving her head. This kind of stylistic excess results in the film's appearing as a mannered, self-conscious exercise in style (albeit a witty and entertaining one). This tends to reduce the significance of the style – the emphasis is on the signifiers rather than the signifieds. Considering that the film was made in 1935 when the stylistic paradigm was nearing its narrowest phase, the comic self-conscious approach can be seen as one way in which such an excessive style could be permitted. In effect, the comedy is extended from the content to the style.

LES MISERABLES

Les Miserables is a very different kind of film, one in which the style seems to be taken very seriously. Directed by Richard Boleslawski, it was made at Twentieth Century under Darryl F. Zanuck and released by United Artists, one month before the merger with Fox. At this time Twentieth Century was a small studio making prestige films, frequently with historical subjects. For *Les Miserables* Zanuck borrowed Gregg Toland from Goldwyn. Even by this time Toland had established himself as a cinematographer of note. The film was very successful, getting into the top twelve box-office draws for 1935 and getting two Academy Award nominations (for Best Picture and for Toland's cinematography).[40]

Like *Viva Villa!*, the most striking visual aspect of *Les Miserables* is the lighting. The three parts of the film (indicated by titles such as 'Thus ended the first phase in the life of Jean Valjean') each progress from lighter to darker set-ups. The first part deals with the trial of Valjean (Fredric March), his imprisonment, release and visit to a bishop's house. Although no part of this is particularly light, there is a noticeable change to a near low-key style for the sequences in the bishop's house. The opening scenes tend to have little, if any, facial shadow. In the second part of the film (in which Valjean has become a successful businessman) the first scenes are in mid-key, with well-lit faces but some background shading. However, from the first meeting between Valjean and Javert (Charles Laughton) the lighting becomes darker and the contrast ratio increases, with strong key lights and fuller background shadows. The climax of this second part is Valjean's escape at night with his adopted daughter Cosette, pursued by Javert. The sequence begins in the bright exteriors of the grounds of the convent in which Cosette (Rochelle Hudson) is being educated, while Valjean works as a gardener. The lighting then progressively darkens as Javert finds Valjean again and they get caught up in riots by revolutionary students in Paris. This part climaxes with another chase, this time in the Paris sewers (an unusual sequence in which the only sound heard is non-diegetic music). Again the film has moved into near low-key and remains there for the closing eight minutes after the sewer chase.

In addition to facial and background shadow, much use is made of silhouettes in the foreground, and well-defined shadows on walls. When Valjean first enters the bishop's house his shadow is seen before he himself appears. When Javert rediscovers Valjean in Paris near the beginning of the third part, the policeman's presence is indicated by his shadow on a wall. Silhouettes are sometimes used to add to the cross symbolism which appears in the film. As Valjean eats supper with the bishop the window-frame makes a cross in silhouette in the centre of the composition. (Although this is a background silhouette, it is as prominent as many foreground silhouettes.) As Valjean goes to look for Cosette's lover, Marius (John Beal), who has been injured in the riots, the silhouette of a lamp-post makes a cross in the foreground.

Along with the low level of lighting, there is a fairly frequent use of low-angle shots, usually looking up at figures of threat (such as Valjean in the bishop's house and Javert in many scenes), although these angles are never extreme. Canted angles are also used in the more dramatic moments, such as the chase at the end of the second part, and the riot scenes in the third part. Camera movement is fairly conventional, although there are some longer movements. The opening shot establishes the trial scene in a 42-second take in much of which the camera is moving across the court towards Valjean in the dock. The most notable aspect of camera distance is the use of ECUs, particularly in the scene in which Cosette tells Valjean of her love for Marius. This scene even includes an ECU two-shot, at the point at which Valjean realises he might lose her. Despite these variations, the structuring of scenes is very conventional in mid-thirties terms and subjectivity is constructed unproblematically. Thus the only aspects which go beyond the restricted stylistic paradigm are some of the signifying codes, particularly those concerned with lighting.

To understand how this wider signifying paradigm was possible, three points must be considered. Firstly, the film was a prestige production by a small studio which tried to differentiate its films from those of the majors. Allowing some stylistic variation was one way of doing this. A similar process can be seen in other studios. Goldwyn will be discussed in Chapter 9. John Ford's films at RKO, especially *The Informer* and *Mary of Scotland*, can be seen in the same light. The studio was able to make some prestige productions by inviting Ford in as a guest director and allowing him a great deal of freedom. The resulting stylistic variation became, in a sense, a proof of the films' quality, with the signifying codes of lighting and camera angle connoting, at a second-order level, artistic status. The second point to be made is that *Les Miserables* was based on a classic source – Victor Hugo's novel – and so was guaranteed respectability, which also allowed some stylistic freedom. Thirdly, the film's historical setting was able to defuse the political content (the story of the film is very similar to that of *I am a Fugitive from a Chain Gang*, produced by Zanuck when he was at Warner Brothers), which does, after all, include street riots and student revolutionaries (some of whom are shown as sympathetic characters). The heavy religious theme would have a similar effect (and it allows one of the film's authority figures – the bishop – to be portrayed as a good character). In these ways, space was created for a visual style which went beyond the contemporary paradigms. In ideological terms this can be seen as a situation in which gaps and tensions in the dominant ideology allow variant forces to gain expression. More specifically, the Hollywood majors' hold on the market forced the smaller studios to differentiate their product in some way. One effective way was to go for prestige production (an alternative was to go in the opposite direction and concentrate on, for example, cheap westerns). Prestige production meant trying to achieve artistic status in some way, whether by seriousness of content or by what was seen as an expressive (and therefore artistic) style. However, by its very nature, this tended to subvert elements of the dominant ideology.

CONCLUSION

Broadly speaking then, the transitional period of 1933–35 saw an increasing restraint being exercised in both style and content as Hollywood adapted to the changing circumstances, but a restraint which was not always able to conceal the conflicts and tensions which followed from those circumstances. This also allowed different ideological forces to gain expression in films. The conflicts noted in the above four films include not just tensions within the broad terms of a dominant ideology (as evident in the confusions over how to portray democratic but violent revolutionaries in *Viva Villa!* and *Les Miserables*) but more major conflicts as the professional ideologies of such people as James Wong Howe and James Whale led to the breaking (or at least the pressurising) of stylistic norms and as the more regressive aspects of patriarchal ideology (usually kept less explicit within the dominant ideology) force their own distortions, as in *42nd Street*. By 1936, however, the restrained style had reached a dominance which made such conflicts less likely to be expressed.

7 Screwball Restraint: *The Awful Truth*

There are two well-known groups of films which show the restrained style of the mid-thirties in its classic form – the Fred Astaire/Ginger Rogers musicals (particularly *Top Hat*, *Swing Time*, *Follow the Fleet* and *Shall We Dance*) and the screwball comedies. The former group were all made at the same studio (RKO), produced by the same person (Pandro S. Berman) and employed the same cinematographer (David Abel) so their unity of style is not surprising. The latter group, however, covers a wide range of studios, directors and cinematographers, and so one of the best-known of the screwball comedies – *The Awful Truth* – is a more useful illustration of this style.

THE MAKING OF *THE AWFUL TRUTH*

By the mid-1930s Columbia was at its prewar peak of profitability.[1] Harry Cohn had been both president and head of production since 1932, exercising strongly centralised control over a team of producers who were turning out the westerns (such as those starring Buck Jones, Tim McCoy and Ken Maynard) and other 'B' films which were the financial base of Columbia's operation. Jules White was in charge of the short-subject unit whose best-known products (then as now) were the films of the Three Stooges. Later in the decade (1937), serials were added to this base. Such an output gave stability to the studio but for profits occasional big hits were needed. Frank Capra was the resident provider of these, particularly after the unexpected success of *It Happened One Night* in 1934 which helped to make 1935 Columbia's most profitable year of the decade. However, Capra did not make many films, giving Columbia only five more between 1935 and 1939, and Cohn looked to outside directors to make successful films for him. The procedure was for a director to produce and direct his own film, using an associate producer and technical crew from Columbia and with Cohn always retaining a close grip on the financial strings.

Of the various films made under this system, one of the most successful was *The Awful Truth*. Director Leo McCarey had just left Paramount for whom he had directed such films as *Duck Soup* (1933), *Belle of the Nineties* (1934) and *Ruggles of Red Gap* (1935).[2] He was assigned Everett Riskin as associate producer and a technical crew which included Joseph Walker, Capra's regular cinematographer. The script by Viña Delmar was based on Dwight Taylor's adaptation of a play by

Arthur Richman.[3] As usual, however, the credits are not the whole story. Gerald Weales quotes Irene Dunne as saying of the script, 'Leo McCarey wrote an awful lot of it while we sat on the set and waited.'[4] The cast was headed by three Columbia contract players – Cary Grant, Irene Dunne, and Ralph Bellamy.[5] Although more expensively produced than the 'B' films, *The Awful Truth* did not have a lavish budget (unlike Columbia's big film of 1937 – Capra's *Lost Horizon*). The cast is not large and the sets are mostly indoor and not particularly expensively decorated.

The film was premiered in November 1937 and was fairly successful, although not making the top fifteen grossers of 1937–38. It did well at the Academy Awards, winning Columbia its only non-Capra award of the thirties which McCarey won for Best Director. There were also nominations for Best Picture, Best Actress (Dunne), Best Supporting Actor (Bellamy) and for Viña Delmar's script.[6] A factor in the film's not getting into the top box-office draws may have been the type of comedy. Sophisticated comedy about marriage, set in New York and making fun of Westerners and Southerners, was calculated to appeal to city audiences but not to the small-town and rural audience. Margaret Thorp, in her classic 1939 study *America at the Movies*, gives a number of examples showing how the American audience was geographically differentiated, such as the fact that *They Won't Forget*, a 1937 drama of Southern prejudice, did not show south of Washington.[7] The non-city audience was not sophisticated. 'It is in the small town that tastes are most definitely marked. The subtle, the exotic, the unexpected they do not like at all, and they are frankly annoyed by costume pictures.'[8] *The Awful Truth* would not have been appreciated by this audience, particularly if they identified with the character played by Ralph Bellamy, a slow-spoken rancher from Oklahoma who becomes the butt of much humour (and of course there is Cary Grant's comment about going to Tulsa if Oklahoma City gets to be boring – the joke is lost on someone who sees Tulsa and Oklahoma City as more exciting than New York).

SCREWBALL COMEDY

Discussion of the humour in this film leads on to the notion of the screwball comedy. Although often seen as the quintessential thirties Hollywood film genre, it is not altogether easy to isolate a precise definition of this comic style. Some writers see it as defined essentially by Frank Capra's films of the middle and later thirties – *It Happened One Night*, *Mr Deeds Goes to Town*, *You Can't Take It With You* and *Mr Smith Goes to Washington*. Andrew Bergman, although defining the form in fairly broad terms ('a comedy at once warm and healing, yet off-beat and airy'),[9] plumps for the Capra cycle as the crucial examples. Keith Reader, on the other hand, sees Preston Sturges as the key figure and places the sub-genre in the war years, beginning with *The Great McGinty*.[10] Of the thirties films, he mentions only *Twentieth Century* and *Bringing Up Baby* as predecessors.[11] Stanley Cavell discusses what he calls the 'Hollywood comedy of remarriage'[12] and although he does not explicitly link this with the screwball comedy, the seven films he chooses to write about clearly cover this ground – *It Happened One Night*, *The Awful Truth*, *Bringing Up Baby*, *His Girl Friday*, *The Philadelphia Story*, *The Lady Eve* and *Adam's Rib*. His emphasis on remarriage links up with Andrew Bergman's comment that 'the comic technique of these comedies became a means of unifying what had been splintered and divided'.[13] Byrge and Miller, in the introduction to their comprehensive filmography of the genre, relate this feature to the experience of the Depression audience: 'Screwball comedy generated part of its dramatic interest from subverting historic and contemporary class conflicts, subsuming them into the disarming dialectic of sexual attraction versus sexual tension . . . . The characters' ultimate pairing at narrative's end transmitted a unifying illusion to the audience which was welcome during a time of potential social division.'[14]

The most detailed study to date of the screwball comedy is that of Wes Gehring. Gehring emphasises the conflict between the strong heroine and the weak hero. 'Study of screwball comedy should begin with the realization that the genre satirizes the traditional love story. The more eccentric partner, invariably the woman, usually manages a victory over the less assertive, easily frustrated male. The heroine is often assisted by the fact that only she knows that a "courtship" is going on.'[15] This weak hero is considered to be an anti-hero whose traits define the type of comedy. 'The five antiheroic characteristics – abundant leisure time, childlike nature, urban setting, nonpolitical activity, and frustration – represented a major transition in American humour and served as a foundation of a frequently neglected film genre: screwball comedy.'[16] Gehring excludes Capra's films, quoting in justification Jim Leach's description of them as populist rather than screwball.[17] This, however, is a rather arbitrary move, since Capra's films have been generally accepted as screwball films. It suggests Gehring's adherence to a too-narrow definition of the sub-genre.

A more useful approach is suggested by Andrew Britton in his essay on Cary Grant's comedies. He gives a definition which relates the notion of sophistication with the screwball character, and at the same time links up with Cavell's emphasis on remarriage and Gehring's anti-hero. In discussing *The Awful Truth*, *Bringing Up Baby* and *Holiday*, he writes 'The principles [sic] of all three films is to identify "play" in the sense of recovered infantile polymorphousness (which is, effectively, the meaning of "screwball") with "sophistication", the apogee of cultivated adulthood. The sophisticated couple is the couple whose sexuality is no longer organised by the phallus.'[18] The remarriage aspect appears when these films are structured by a realignment of couples. The 'wrong' pairing is premised at the opening of the film and forms the disruption which sets the narrative in progress. The films end when the two screwball characters are brought together in an alignment which suggests an ideal pairing. *It Happened One Night*, *Bringing Up Baby*, *Holiday*, *His Girl Friday* and *The Lady Eve* all adhere to this pattern. *The Awful Truth* is simply a variant of this with the screwball couple breaking up at the beginning and pairing off with non-screwball characters before the final reconciliation (a pattern also found in *The Philadelphia Story*). This allows the remarriage motif to be seen simply as a narrative device to show up the nature of the screwball character by contrast with a conventional character and not as a necessary or defining element of the sub-genre. Broadened in this way, *Mr Deeds Goes to Town*, *Nothing Sacred*, *Easy Living*, *You Can't Take It With You* and *Mr Smith Goes to Washington* can be included in the sub-genre, these being films in which the screwball couple ends up together even if there are no other potential partners with which to contrast them. Britton's emphasis on play is useful as a means of defining the screwball character, although his notion of the screwball couple as 'the couple whose sexuality is no longer organised by the phallus' is rather too narrow. He himself admits that *It Happened One Night* does not fit into this definition – but neither do, for example, *Nothing Sacred*, *His Girl Friday* and *Mr Smith Goes to Washington* (nor do these films feature Gehring's anti-hero).

A history of the screwball comedy can now be outlined, to show the position of *The Awful Truth* in the sub-genre's development. The origin can be seen in two types of comedy in the earlier thirties – the sophisticated comedies of Ernst Lubitsch (*Trouble in Paradise*, *Design for Living*, *One Hour With You*) and the verbally and physically fast comedies of the Marx Brothers. Screwball comedy is sometimes seen as quintessentially verbal comedy but it must be remembered that there is a strong element of physical, even slapstick, humour in it, along with frequent reliance on props and animals. Some screwball comedies tilt over into pure visual humour for major sequences, for example *Easy Living* and *Topper*. As Lewis Jacobs points out, 'these films were all sophisticated, mature, full of violence – hitting, falling, throwing, acrobatics – bright dialogue, slapstick action – all imbued with terrific energy'.[19]

The sub-genre develops during 1934 – the year of *Twentieth Century* (Columbia, d. Howard Hawks) *It Happened One Night* (Columbia, d. Frank Capra) and *The Thin Man* (MGM, d. W. S. Van Dyke). (Although ostensibly a detective mystery, this last film has a central partnership – played by William Powell and Myrna Loy – who come very close to the classic screwball couple.) In 1935 come *Ruggles of Red Gap* (Paramount, d. Leo McCarey) and *Hands Across the Table* (Paramount, d. Mitchell Leisen), both films which are very near to the classic screwball form, especially the latter. This is then followed by the full appearance of the sub-genre in 1936. The following list of some of the principal examples will show the chronological development in the later thirties, as well as indicating the studios and directors most involved.

1936	April	*Mr Deeds Goes to Town* (Columbia, d. Frank Capra)
	September	*My Man Godfrey* (Universal, d. Gregory La Cava)
1937	June	*Easy Living* (Paramount, d. Mitchell Leisen)
	August	*Topper* (MGM, d. Norman Z. McLeod)
	November	*Nothing Sacred* (Selznick, d. William A. Wellman)
		The Awful Truth (Columbia, d. Leo McCarey)
	December	*True Confession* (Paramount, d. Wesley Ruggles)
1938	March	*Bringing Up Baby* (Columbia, d. Howard Hawks)

	June	*Holiday* (Columbia, d. George Cukor)
	September	*You Can't Take It With You* (Columbia, d. Frank Capra)
1939	April	*Midnight* (Paramount, d. Mitchell Leisen)
	October	*Mr Smith Goes to Washington* (Columbia, d. Frank Capra)
1940	January	*His Girl Friday* (Columbia, d. Howard Hawks)
	December	*The Philadelphia Story* (MGM, d. George Cukor)

Parallel to this list, a second one could be made of films with some screwball characters, although not examples of the classic screwball couple, for example *Young in Heart* (Selznick, d. Richard Wallace) and *Topper Takes a Trip* (Hal Roach, d. Norman Z. McLeod), both from 1938. As Reader has noted, the sub-genre continues during the early forties with the films of Preston Sturges, such as *The Great McGinty* (1940), *Sullivan's Travels* (1941), *The Lady Eve* (1941), *The Miracle of Morgan's Creek* (1943) and *Hail the Conquering Hero* (1944). It is also worth noting that some musicals, particularly those of Fred Astaire and Ginger Rogers, get very close to the classic screwball conventions.

In this chronology, then, *The Awful Truth* can be seen as central to the first full development of the screwball comedy, and Columbia's role can also be clearly seen.

THE SCRIPT

Jerry Warriner (played by Cary Grant) returns home pretending to his wife Lucy (Irene Dunne) that he has been on holiday in Florida. However, he finds that she is not at home, appearing later with her French singing teacher Armand Duvalle (Alexander D'Arcy), telling how their car broke down and they had to spend the night in a small hotel in the country. Jerry's lack of trust results in the couple deciding to get a divorce. The court gives them ninety days before their divorce becomes final and Lucy goes to live with her Aunt Patsy (Cecil Cunningham). She meets rich Oklahoma rancher Daniel Leeson (Ralph Bellamy) but her engagement to him is ended when Jerry and Armand are found fighting in her bedroom. Jerry then has a romance with rich heiress Barbara Vance (Molly Lamont) but this is also ended – this time by Lucy pretending to be Jerry's drunken sister Lola. On the night before their divorce becomes final Lucy engineers their seclusion in Aunt Patsy's remote Connecticut cabin. Jerry finally agrees to be more trustful and they are reconciled a few minutes before midnight.

What is only partially clear from this synopsis is how the central couple are defined by oppositions. As a sophisticated city couple they are, in the first half of the film, opposed to Daniel, the slow-speaking naive Westerner (whose final words, after seeing Jerry and Armand run through Lucy's apartment, are 'Well, I guess a man's best friend is his mother'), and to Dixie Belle (Joyce Compton), a night club singer Jerry meets (who uses a Southern accent for career purposes and whose act is unsophisticated enough to include a wind blowing up under her dress each time she sings the line 'My dreams have gone with the wind'). In the second half of the film the opposition is provided by the rich blue-blood family of the Vances, with their insistence on social status. What they object to is not so much 'Lola's' drinking but what she says about Jerry's lowly background (the father whom Jerry claimed went to Princeton, Lola says was a gardener there).

The Awful Truth works to establish a particular norm for married couples, involving absolute trust and no privileges based on sexual difference. From one point of view this is simply an idealisation of a traditional view of marriage. The Warriners' financial independence – there is no mention of them having to work and their house is a large one – can be seen as another aspect of this process. However, the film allows for another reading. Although the Warriners are clearly well-off, their house only appears in one scene. They establish themselves as a couple who are complete in themselves, not needing the support of property as do Daniel and Barbara. Their lack of interest in children suggests a view of marriage rather different from the traditional one in which reproduction within the family is central to the social structure. Rather than setting up a nuclear family unit, Jerry and Lucy simply want to enjoy each other's company and have fun. This could be read as an ideologically conservative mystification of the real role of the institution of marriage. However the absolute trust which Lucy demands can function as a critique of this role. This trust demands a rejection of conventional double standards and, more importantly, male domination (and here the relevance of Britton's comments, mentioned earlier, are clear). If marriage is seen as a microcosm of society, then Lucy's demands suggest a social revolution.

This aspect of the film, however, is not prominent enough to ensure the rejection of the more conventional reading. As with most of these comedies, the screwball characters can be taken merely as jolly eccentrics, showing what fun it is to be one of the idle rich, or as representatives of more fundamental criticisms of the repressions of society. If cinematographic style has an ideological role to play, then it should be apparent in a case like this. In the middle of Hollywood's most restrained period – the later mid-thirties – *The Awful Truth* seems to allow for a socially deviant reading. How cinematographic style affects this could be crucial.

CAMERA MOVEMENT

Over two 100-shot sequences (out of the film's total of 705 shots), the statistics for camera movement were as follows.

Static camera	70.5%
Reframing movements	12.5%
Pans and tilts	10.5%
Tracking shots	6.5%

Here is the full domination of the reframing style – more than three-quarters of the shots are either static or contain very small movements. Even the larger movements are limited almost entirely to following a character from one part of a room to another. The nearest the film gets to a camera movement independent of any character movement is in the opening sequence. The first four shots are as follows.

1 Stock shot of a tug on New York's East River.
2 Low-angle static shot of a clock striking eight.
3 The camera moves in towards a sign reading 'Gotham Athletic Club'.
4 Dissolve to Jerry inside the club.

The third shot – lasting only three seconds – is the least-motivated camera movement in the film and yet even it is limited in its significance, acting as a conventional opening, bringing the viewer into the narrative. The very brevity of the shot restricts it, even within the conventional context (*Top Hat* begins with a similar club sign but the first three shots, which move into the club and up to Astaire sitting in a chair, include one 24-second movement and one of 39 seconds). Another movement that may seem to be unmotivated is during the 'bowler hat' scene. Lucy hides Armand's bowler hat behind a mirror but her dog ('Mr Smith') thinks she is playing a game with him. The camera follows Lucy as she hides the hat and then moves down and forward to a closer view of the dog. The shot lasts for seven seconds. The movement is unmotivated in the sense that neither Jerry nor Lucy is looking in that direction at that moment. However, the movement can be seen as linking one character's point of view (Lucy's) with another's (the dog's) and so as having a weak motivation.

Other forms of conventional camera movement common at this time do not appear, or only in a very weak form. The parallel tracking shot, following a character as he or she walks, is present only in very short examples, such as a movement across a hallway (as in the shot in which Aunt Patsy first meets Daniel), or as a faked movement (as in the shot of Jerry and Lucy riding on the traffic cops' bike – the only real movement is in the back projection). Similarly the 'survey' tracking shot (an extended track or crane movement looking over a scene and acting as a mobile establishing shot) is also present only in a very restricted form. An example is at the opening of the court scene. The camera begins with a view of the judge with two clerks sitting in front of him and then slowly moves back revealing the rest of the courtroom. The shot lasts for 19 seconds but the camera stops moving after 10 seconds. Most scenes start simply with a pan or a static establishing shot, the pans sometimes including a small tracking movement as well.

From these examples it can be seen that the camera movement in *The Awful Truth* is highly restrained, even within the conventions of the period, being little more than reframing.

CAMERA DISTANCE

The statistics for distance over the two 100-shot sequences are as follows.

CU	2%
MCU	15.5%
MS	22%
MLS	41.5%
LS	10%
ELS	9%

These statistics are slightly misleading in two ways. There are no CUs of people anywhere in the film. The 2 per cent noted above are shots of the dog. The percentage for ELSs is higher than would have been found if the whole film had been assessed since one of the sample sequences includes the montage showing Jerry and Barbara at various sporting events. There are more ELSs in this montage sequence than in the rest of the film put together. The only other ELSs are conventional establishing shots (for example, the opening shot of the East River and the high-angle shot of the nightclub dance-floor at the beginning of the 'Gone With the Wind' scene). These statistics show why French film critics have labelled the MLS as *le plan americain* – it is totally dominant. Almost two-thirds of the shots are either at that distance or at MS.

Within scenes, camera distances are structured very predictably. Establishing shots are at LS or ELS. Re-establishing shots (shots within scenes which re-establish the viewer's sense of the scene's geography and which allow for major movements across sets to be clearly understood) are at LS. The more important reaction shots and shots in which a detail of facial expression is important are at MCU. The rest are in the MS–MLS range, usually alternating between the two to give a sense of variation. *The Awful Truth* sticks to this structure throughout. A clear example is the first scene of Lucy and Aunt Patsy in their apartment. It begins with Lucy alone with her aunt. At shot 11 her aunt meets Daniel in the hallway outside and invites him in. At shot 30 Jerry arrives to visit the dog. The relationship between camera distance and action will be seen in the following table.

Shot	*Distance*	*Action*
1	LS	Camera pans across room with Aunt Patsy.
2	MLS	
3	MS	
4	MS	
5	MLS	
6	MS	Lucy and Aunt Patsy talk.
7	MS	
8	MS	
9	MLS	
10	LS	Aunt Patsy leaves the room.

Shot	Type	Description
11	MS–LS–MLS	Patsy meets Daniel in the hall (65 sec. take).
12	MS	
13	MLS	
14	MS	
15	MLS	
16	MS	
17	MCU	
18	MS	
19	MCU	Lucy, Aunt Patsy and Daniel talk. 17, 19 and 21 are closer shots of Aunt Patsy as she recommends Daniel to Lucy with meaningful eyebrow-raising.
20	MS	
21	MCU	
22	MLS	
23	MS	
24	MS	
25	MS	
26	MS	
27	MS	
28	MS	
29	MLS	
30	MS–LS–MLS	Jerry enters the room.
31	MLS	
32	MLS	Lucy introduces Jerry to Daniel.
33	MLS	
34	LS	Jerry moves over to the piano.
35	MLS	
36	MLS	
37	MLS	
38	MLS	
39	MLS	
40	MLS	
41	MLS	Lucy, Daniel and Aunt Patsy try to talk while Jerry plays the piano with the dog.
42	MLS	
43	MLS	
44	MLS	
45	MLS	
46	MLS	
47	MLS	
48	MS	
49	MLS	
50	LS	50 and 52 are longer shots as Jerry wrestles with the dog on the floor.
51	MLS	
52	LS	
53	MLS	
54	MLS	

55	LS	Lucy, Aunt Patsy and Daniel leave the room.
56	MLS	Jerry reacts to Lucy's farewell gesture.
57	MS–LS–MLS	Lucy, Aunt Patsy and Daniel in the hallway.

The point of calling structures like this predictable is not that before seeing the film the viewer can predict exactly what the distances will be, but that given the script, the general pattern of distances will be predictable.

CAMERA ANGLE

In the sample, 55 per cent of shots were from a camera shooting parallel to the ground. 30 per cent were slight high angles motivated by the action (for example, looking down at a character in a chair). Eight per cent were motivated high angles; 1.5 per cent were motivated low or slight low angles. This leaves only 5.5 per cent as unmotivated angles, most of which were slight low angles. There are no extreme angles of any kind in the film. The high number of motivated angles (in comparison with the unmotivated angles) shows that generally angles are used to reinforce the viewer's sense of where people are in the set, although even here some sense of the social relationships is conveyed since the people who stand and the people who sit are usually carefully chosen. In the 73 shots of the scene in the Vances' house, there is a sequence of 22 slight high angles (out of a total of 38 such angles in the whole scene). These 22 shots are of four characters sitting down (Jerry, Barbara, Mrs Vance and Lucy as 'Lola'). The shots are given a strong motivation by the fact that Mr Vance is standing throughout the scene. That there are far more high angles than low angles in the film as a whole is a reflection of the fact that in this type of comedy characters' vulnerability is emphasised, rather than their power.

Another detail of camera angle worth noting is the use of near direct-to-camera shots in the 'Gone With the Wind' nightclub scene. In the final part of the scene Daniel and Lucy are dancing energetically to up-tempo music, much to Lucy's embarrassment. Jerry is watching from a side-table, enjoying the spectacle. Six of the shots of Jerry are very close to direct looks at the camera. However in four of them he is looking very slightly below the camera (the dance floor is a few feet lower than the level on which the surrounding tables are placed) and in the other two he is looking very slightly to the side of the camera. None of these shots is emphasised in any way and they are all tied in with reverse angles of Daniel and Lucy dancing (in some of which Lucy glances back at Jerry).

LIGHTING

The lighting in *The Awful Truth* is almost entirely high-key, with no use of strong shadows and very little additional star lighting. Only in the final scenes (after the visit to the Vances) are there any darker scenes. These final scenes isolate the central couple as they are reunited. Thus the darkness has conventional romantic connotations (although the approach of midnight and the threat of the divorce becoming final also allows some connotation of danger and isolation in the darkness). There is no use of darkness anywhere else in the film and even in these final scenes there are no strongly defined facial shadows (although the use of wall shadow is more marked). It is worth commenting on the high-key lighting, if only to point out how unnatural it is. By flooding the set in light, even small details in the corners of rooms are fully lit. This puts up the cost of production, both by using greater amounts of electricity and by necessitating more detailed set decoration. In addition it acts against conventional star lighting in that it is difficult to glamorise a star in extra lighting if the general level of light is already very high. (It is notable that Garbo's films of this period – *Anna Karenina* (1935), *Camille* (1936), *Conquest* (1937) – are lit in a more subdued tone than full high-key, allowing the full effect of her own star lighting to be seen). Thus there are various commercial arguments against using full high-key lighting. This is simply to point out that its use in a film such as *The Awful Truth* is not simply an automatic choice without financial implications.

One other scene is worth mentioning in this context. Although of minor importance in itself, it does emphasise the lighting elsewhere in the film. This is the scene in which Lucy and Aunt Patsy read in a newspaper about Jerry's forthcoming engagement to Barbara. It is a brief scene of four shots, lasting only 47 seconds.

1. Lucy is sitting reading her newspaper, face completely in shadow. She gets up and walks over to Aunt Patsy sitting on her bed. The camera pans with her.
2. Close-up of a newspaper picture of Jerry and Barbara.
3. Lucy and Aunt Patsy have their backs to the camera. Lucy stands up and turns round, sitting down again, facing Aunt Patsy.
4. Close-up of the caption under the newspaper picture.

This is immediately followed by the montage sequence of Jerry and Barbara at sporting events. There are two oddities in the lighting here. The first is the complete facial shadow at the beginning which is completely out of character with the rest of the film but which is too unemphasised to count as an important part of the filmic system. The second oddity is that when Lucy stands up in the third shot, the lighting (which has been planned for her sitting down) becomes too strong on her shoulders and back, giving ultra-strong top and back lighting – again an effect which is completely out of character with the rest of the film. The likely explanation of this is that the scene was found necessary as an introduction to the montage sequence but that this discovery was made only after the shooting had been completed and so a different director and cinematographer were used. The scene is of very minor importance and the lighting oddities do not stand out strongly, but they are worth mentioning simply to underline how standard is the lighting in the rest of the film.

SIGNIFICATION

The sub-codes at work in *The Awful Truth* are extremely restrained. The ideological importance of this is that they work to limit the possibilities of meaning. The elements which the sub-codes eliminate are precisely those elements which tend towards ambiguity – unmotivated angles, extended camera movements, high-contrast lighting, extreme differences of camera distance. The openness and clarity of the lighting lays all before it. The deceptions in the narrative are between characters, not between the viewer and the text. The placings of the camera – movement, distance and angle – all serve to convey the narrative with as little addition to the meaning as possible. Such a style will tend to pass on the ideological assumptions of the narrative content with as little distortion (or even emphasis) as possible. The connotations of openness (lighting), stability (movement), equality between viewer and character (angle) and 'correctness' of distance (neither too far away to distance the characters nor yet too near to create too strong emotional identifications) all function to suggest an essentially non-problematic text, with an obviousness of viewer identification.

The ambivalence of the content of *The Awful Truth* has already been mentioned. At first sight it might seem that this 'open' style would pass on any such ambivalence. This, however, is not the case. The style works against ambivalence (this is the precise connotation of high-key lighting) and against any instability or even subtlety. Since the reading of *The Awful Truth* which emphasises the deviance of its assumptions is the more subtle reading, the style works against this. The 'open' style tilts interpretation towards the more obvious reading, seeing the film as 'just a comedy', romanticising married love.

As already noted, the structuring of signifying elements is very predictable. Unlike that of *All Quiet on the Western Front* and *Dr Jekyll and Mr Hyde*, it does not draw attention to itself but yet forms very strong scenic structures. Its predictability is itself a limiting feature, suggesting that the film is essentially like any other film. By having no disruptive structures, no structures which vary the norm in unpredictable ways, a container is made into which almost any content can be poured without disrupting the overall meaning.

To emphasise just how undisturbing the structuring is, two points can be made. The first concerns the flattening of the arc structure. In Chapter 2 the basic arc structure was described as a structure which produces an arc if camera distance is plotted on a graph against time, with the camera moving from long establishing shots into the middle distance and then close for the climax, before moving out again. It was also noted that the arc is frequently distorted by the move out at the end being quicker than the move in during the first part of a scene, giving a lop-sided arc. In *The Awful Truth* a further distortion takes place. The general lack of extreme distances in the film (particularly the ELS and the CU) results in a flattening of the lop-sided arc. Thus variation within the scene is reduced, making the structuring all the more stable and predictable.

The second point to be made about the structuring is that scenes are frequently quite long. Thus not only is the arc flattened but it is also elongated. This process is evident in the scene at Aunt Patsy's apartment described above. When different segments of a scene are linked only by reframing movements, the final outcome is a structuring which is barely noticeable, thus becoming the ideal non-intrusive stylistic device. The structure signifies by absence.

Adding to these various effects of signification and related to them there is another level of connotation in this classic film style, much like Barthes' notion of myth. The open, seemingly unproblematic style has in itself an overall connotation – the myth of Hollywood. In this myth Hollywood is just the producer of entertainment with no political or social axes to grind. As such it is part of the broader myth of the USA as the land of liberty, opportunity and the classless society – at which point the overlap of mythology and ideology is complete. This second-order connotation is arguably present to some extent in all those films which are conventionally identified as 'Hollywood films'. However those films, like *The Awful Truth*, which are made in this seemingly obvious style of high-key lighting, reframing movement and non-intrusive structuring, carry this connotation in a much stronger fashion. The match between the 'open' style of the films and the 'open' myth of Hollywood is perfectly made.

SUBJECTIVITY

Point-of-view in *The Awful Truth* is organised as a conflict between the two central characters. The opening scenes in the club and in the Warriners' house are centred on Jerry's point-of-view. That is to say, shots are either centred on Jerry (showing him looking at and addressing other characters) or represent what he is looking at (not from an exact point-of-view but from a 'third person' position). In other words, his subjectivity is central to the organisation of camera point-of-view, although this camera positioning coheres around the point-of-view of an absent third person. That third person is, as it were, led by Jerry's point-of-view. Opposed to these scenes are those centred on Lucy, such as the scenes in her apartment. Most scenes in the film fit into one of these two opposing point-of-view structures, with the final scene more equally divided between them. Even the montage sequence of Jerry and Barbara visiting and taking part in sporting events is organised around Jerry's point-of-view (unlike the typical thirties montage sequence which tends to break the point-of-view structure by departing from any direct relationship with the main characters). A few brief scenes develop other points-of-view but they are always closely related to either Jerry or Lucy. For example, the two shots of Daniel with his mother are part of what Metz would call an alternating syntagm with two shots of Lucy and Aunt Patsy. Similarly the single shot of the lawyer on the phone to Lucy can almost be regarded as Lucy's 'point-of-hearing', rather than her point-of-view. The only other type of break in the point-of-view structure is the establishing shot, such as the opening two shots of the film, described earlier. But these, as with the montage sequences, are very conventionalised breaks in point-of-view and thus do not disrupt the subject position. In fact they can be said to strengthen it since they serve to introduce and interpellate the subject position, gently manipulating the viewer's point-of-view.

What all this results in is a film which very strongly seeks to position the viewer. The combination of point-of-view, strong structuring and fairly subtle interpellation (there are no strong points of address in the film after the titles, merely a continual flow of unified imagery) constructs an apparatus which aims at never leaving the viewer in any doubt as to his or her relationship with the text.

CONCLUSION

As was argued earlier, of the two systems of signification and subjectivity, the latter is potentially the stronger. The strength of the system of subjectivity in *The Awful Truth*, along with the working of the system of signification (including the structuring of signifying elements) leads to the conclusion that although it can be read as an ideologically subversive film, the cinematographic style is continually playing against such a reading. In this it is typical of the screwball comedy. As a group these films contain some of the strongest social criticism to be found in Hollywood films of the middle and later thirties, notably in *My Man Godfrey*, *Nothing Sacred*, *Holiday* and *Easy Living*. Yet the style of these films consistently contradicts this element. It is not surprising to find that stylistically the screwball comedy is the most conservative of all genres in the thirties. It makes an interesting contrast with the horror film. The classic horror films work by releasing a repressed element and then violently repressing it when it gets out of hand. In content it is thus potentially the most conservative of all genres. Yet in style – at least in the thirties – it is the freest genre, and this is true not just for the early examples from 1931–33 but also for those of the mid-thirties such as *Bride of Frankenstein* (as already noted in Chapter 6) and *Dracula's Daughter*. The variety of cinematography found in these films is far greater than anything found in the screwball comedies.

Such considerations should make it clear that the emergence of the screwball comedy at precisely the time at which film style was becoming most restrained was not accidental. The screwball comedy is not just a comedy of manners or a comedy of the idle rich. It is a potentially subversive form which can easily carry a strong critical force. The restrained style of these films is not merely the accompaniment of the humour but rather the condition of the humour's existence. Only with a 'safe' style which offers the maximum power of containment could these films give expression to their particular form of comedy in the mid-thirties.

As a final comment on *The Awful Truth* it is worthwhile noting that despite the restrained cinematographic style, it is still possible to read the film as subversive, thus showing the limitations of style as a determining feature. As with other ideologically influenced elements, style is just one aspect among many and even when all the elements are pulling together they still need to contend with the reader of the text, situated in his or her own ideological setting. What this analysis shows is the pattern of forces at work within the text.

8 The Restrained Style, 1936–38

By 1936 what Bordwell, Staiger and Thompson have labelled 'the classical paradigm' had reached its most restrained and most restricted version. As the government attempted (none too successfully) to pull the country out of recession, and the more extreme signs of crisis faded, Hollywood (as noted in Chapter 6) reacted in its own way. The Production Code was by now making its full impact on Hollywood scripts, and film content had changed in various ways, most obviously by featuring a plethora of child stars. Shirley Temple had been a star for Fox since 1934 but 1936 saw Judy Garland's first screen appearance for MGM and Deanna Durbin making her first starring feature for Universal, released early in 1937. Later in 1937 came *A Family Affair*, the first of the MGM Andy Hardy films with Garland and Mickey Rooney. Other child stars included Freddie Bartholomew (*David Copperfield*, 1935, *Little Lord Fauntleroy*, 1935, *Captains Courageous*, 1937, *Kidnapped*, 1938) and Jackie Cooper (whose career had started at the end of the silent era and was a well-known face throughout the thirties). Presenting a rather different image of childhood were the Dead End Kids (see the following chapter). If the increasing importance of child stars was the most obvious consequences of the Production Code's restrictions, it was certainly not the only one. James Cagney and Edward G. Robinson found themselves playing characters on the side of law enforcement rather than the gangster roles which had first brought them fame. Typical musicals were the RKO Astaire/Rogers films and the MGM operetta-style musicals of Jeanette MacDonald and Nelson Eddy, rather than the topical backstage musicals which followed *42nd Street*. Mae West had to tone down her films and her career took a dive. Historical melodramas gained a new popularity with the studios since more revealing dresses were allowed than in contemporary topics. However in order to appreciate fully the changes in film content and style at this time, it is important to note the changes that were going on in the studios themselves.

THE STUDIOS IN THE MID-THIRTIES

As the transitional period of 1933–35 ended, a series of important management changes took place in the Hollywood studios.[1] These changes are worth noting for two reasons. They add force to the claim that the studio system during the thirties was not the stable economic structure which it is often portrayed as, and more importantly for the present purposes, the changes in many cases resulted in more public changes in the film output itself.

The main changes can be listed as follows.

1935 Lubitsch and Herzbrun take charge of production at Paramount.
Twentieth Century merges with Fox
Joseph Schenck leaves United Artists.
Selznick joins United Artists.
Atlas Corporation buys into RKO.
Republic and Monogram are merged by Herbert Yates.
1936 A new Monogram is launched, independent of Republic.
Balaban takes over from Lubitsch and Herzbrun at Paramount.
Irving Thalberg dies.
Carl Laemmle, Sr is bought out of control at Universal.
Walter Wanger joins United Artists.

The implications of these changes may not be immediately apparent. The antipathy between Lubitsch and Sternberg meant that with Lubitsch's accession, Sternberg's days at Paramount were numbered.[2] At this time the experimentation evident at Paramount in the earlier thirties, particularly in the work of Sternberg and Mamoulian, as well as the so-called 'European' sensibility (which mainly referred to what was seen as sexual sophistication) most associated with Lubitsch's own work, all but disappeared. The death of Thalberg (as well as causing what Baxter calls a 'general confusion at Metro')[3] consolidated Mayer's position. Instead of Thalberg's role as what Gomery calls a 'super-producer'[4] in charge of MGM's most prestigious productions (*Romeo and Juliet* being the last, treated as his memorial), Mayer had full control of a number of producers, each with less power than Thalberg had had. As Mayer took full charge, the prestige films of Thalberg (mainly versions of literary classics) began to disappear from the MGM production schedules.

Laemmle's loss of control marked a change in Universal's output. Before 1936, it was best-known for horror films such as *Dracula* (1931), *Frankenstein* (1931), *The Old Dark House* (1932), *The Mummy* (1933), *The Black Cat* (1934), *The Werewolf of London* (1935), *Bride of Frankenstein* (1935) and *The Raven* (1935). In 1936 the emphasis changed to musicals, including *Show Boat* in that year (a transitional film in which director James Whale brought his rathered mannered style to bear on a popular musical) and then the Deanna Durbin films – *Three Smart Girls* (1937), *One Hundred Men and a Girl* (1937), *Mad About Music* (1938) and *That Certain Age* (1938).

Twentieth Century's take-over of Fox also resulted in a changed product. Twentieth Century's prestige productions (such as *The Bowery*, 1933, *House of Rothschild*, 1934, and *Les Miserables*, 1935) provided a much-needed boost to Fox's output and the post-1935 films include both Shirley Temple musicals and more respectable historical subjects such as those directed by Henry King (*Lloyds of London*, 1936, *In Old Chicago*, 1938, *Stanley and Livingstone*, 1939) and John Ford (*The Prisoner of Shark Island*, 1936, *Four Men and a Prayer*, 1938, *Young Mr Lincoln*, 1939, *Drums Along the Mohawk*, 1939).

Even Warner Brothers, although not undergoing important management changes, also changed its image in 1935, with *Captain Blood* making the move away from the topical films such as the gangster films and the backstage musicals, to period films, in particular Errol Flynn swashbucklers and Paul Muni biopics. To sum up, the sorts of films the studios were famous for in 1936 were different from the sorts of films they had been famous for in the early thirties.

These studio changes were, of course, closely related to the economic situation. In particular, the changes at RKO, Universal and Fox were directly linked to those studios' problems during the first part of the Depression. RKO went into receivership in 1933, between 1932 and 1938 Universal only once managed to make an annual profit, and as for Fox, Gomery comments that 'in 1931 Fox was floundering in a financial morass'.[5] Even the changes in output at Warner Brothers and MGM were motivated by the need to attract the largest possible audiences, which for MGM meant abandoning its most up-market films and for Warner Brothers meant expanding out of the limited market it had already captured.

Returning to the content of these films, the major contrast with the early thirties is that where the earlier period had concentrated on narratives of survival, by the 1936–38 period the dominant narrative has become one of relatively unproblematic celebration – celebration of youth, of history, of traditional gender roles and even of traditional authority figures. In these narratives personal tragedies become celebrations of the human spirit (the Garbo vehicle *Camille* being a famous example which manages to romanticise and celebrate death by illness) and even historical disasters become by implication celebrations of what later generations will achieve (as in the ending of Twentieth-Century Fox's account of the great Chicago fire of 1871, *In Old Chicago*). If the screwball comedies, as noted in the previous chapter, were able to suggest some subversions of these attitudes, it was only within a celebratory narrative of the idiosyncrasies of human nature. The change is perhaps best summed up in the change in the Marx Brothers films when they moved from Paramount to MGM in 1935. The Paramount films are anarchic, attacking all and sundry with an at times fairly vicious humour, most famously seen in the war sequence at the end of *Duck Soup*. The MGM films (starting with *A Night at the Opera* in 1935) are, however, much closer to conventional comedy, putting more emphasis on verbal puns and playing up the romantic and musical aspects of their plots.

If the various changes can be linked by a common desire of the studios to find a safe product, then it should not be surprising that 1936–38 is the period when the stylistic paradigm was at its most restrained. *The Awful Truth* has already provided an account of this restrained style, with its predictable structuring, unproblematic subject positioning and signifiers of stability, equality, clarity and openness. This style becomes the visual equivalent of the restrictions on content which the 1934 enforcement of the Production Code marks. Not only does the mid-thirties style retreat from the extended paradigm of the early thirties, but it also retrospectively signifies that earlier style as excessive. If this, the style of 1936–38, is promoted as normality (and the ease with which many writers even now identify this style with the style of traditional Hollywood suggests that this promotion is still successful) then that style which went before must be deviant. The high-key, reframing, continuity style signifies the dismissal of the experimental extremes of the earlier years. The later style proclaims its naturalness (as does all ideology) by its very obviousness. What could be more obvious than high-key lighting, reframing movements and eye-level camera positions? The style seems to lay all before it, concealing nothing.

To add detail to this account, two films will be considered which, when taken along with *The Awful Truth*, will show the range of styles available at this time. The first contrasts with the screwball comedy by being a costume film with many outdoor scenes, as well as being shot in colour – *The Adventures of Robin Hood*. The second is the kind of film which might be thought to be a prime candidate for a more expressive style. It is a tragic, romantic melodrama with political overtones – *Three Comrades*. Both films date from 1938.

THE ADVENTURES OF ROBIN HOOD

Made at Warner Brothers and produced by Hal B. Wallis, *The Adventures of Robin Hood* involved the participation of four directors.[6] Production began with William Keighley who shot some of the forest scenes but, being said to be too slow, was replaced by Michael Curtiz. B. Reeves Eason was second-unit director whose contributions included part of the forest ambush sequence. Finally, William Dieterle shot a very brief sequence. Tony Gaudio began as cinematographer but was replaced by Sol Polito when Curtiz took over direction. Because the film was shot in colour an advisory team from Technicolor was on the set (this was insisted upon by Technicolor for all productions using their process).[7] It consisted of Natalie Kalmus and cinematographer W. Howard Greene. At a cost of almost $2 million, *The Adventures of Robin Hood* was Warners' most expensive film at the time. However, it was a great success, appearing in the top fifteen box office draws of 1937–38. It won Academy Awards for Erich Wolfgang Korngold's music, the art direction of Carl Jules Weyl and for the editing of Ralph Dawson, as well as a nomination for Best Picture.[8]

Stylistically the film is interesting because, although it is made in the same restrained style as *The Awful Truth*, it shows the outer limits of that style. Most of the lighting is high-key, camera movements almost always follow character movement, there are no ECUs, scenes are structured in predictable patterns and subjectivity is carefully constructed with even the ELSs placing the viewer among groups of observers. Despite this there are interesting, if minor, exceptions – points at which the stylistic limits are reached.

Although most of the lighting is high-key, there are a few short scenes in which darker tones are used. The first is when Robin (Errol Flynn), Much (Herbert Mundin) and Will Scarlett (Patric Knowles) escape on horseback from Nottingham Castle, pursued by Norman soldiers. The sequence is shot 'day-for-night' using filters. Despite this, the trees of the forest through which they ride are still able to produce large dark shadows. One shot includes strongly directional rays of light crossing the frame at an angle of 45° to the horizontal. Another shows a horseman riding through a river, with the dark surroundings contrasting with strong light reflected by the splashing water. The day-for-night technique results in some shots looking like genuine night shots and others looking like shots taken in a dense forest in late afternoon sunlight, but the point is that some of these shots do include very strong shadows.

Other uses of shadow are even briefer. The escape sequence is followed by a scene in which Prince John (Claude Rains) declares Robin an outlaw. It begins with a shot of two shadows on a wall before the camera moves over to reveal Prince John, Sir Guy of Gisbourne (Basil Rathbone) and two other Norman nobles. Soon after this there is a short sequence illustrating the Normans' reign of terror and which includes the most visually striking scene in the film. A Saxon is about to be hanged. He and his Norman captors are in near silhouette. The background is an intense, orange-red sunset. This four-shot scene includes two MSs of a Norman on horseback with his eyes completely shaded. The use of colour in this brief scene is more typical of Technicolor romantic melodramas such as *The Garden of Allah* and *Gone With the Wind*, both of which were Selznick productions, rather than mainstream studio films. Another dark four-shot scene occurs after the archery contest as Robin is thrown into a dungeon. There is heavy background shadow throughout and the scene begins with wall shadows before the sources of the shadows are seen. The fourth shot shows Robin leaning against a wall and his captor, who is tying him up, is seen only as a shadow.

The final dark scenes are even briefer. After Robin escapes, he returns to visit Marian (Olivia de Havilland). The scene of their meeting is fairly well lit – even though the backgrounds are darkish, the two characters are shown in bright light – but the scene opens with two exterior shots, as Robin climbs up the castle wall to Marian's window. Both shots are very dark. This exterior view is repeated in one shot in the middle of the scene and in three shots at the end. These six shots frame 39 interior shots, all much more brightly lit. Dark background is also found in the opening ELS of Marian's trial scene. She is in white and very brightly lit, at the centre of the frame. The background walls and guards are much darker, with deep shadow. Finally, in the climactic duel between Robin and Sir Guy, fighting wall shadows appear several times.

Three points can be made about this lighting scheme. Firstly, all these darker set-ups are in brief scenes or are brief sections of longer, brighter scenes. The overwhelmingly dominant style is high-key. Secondly, there is virtually no use of facial shadow, except in the sunset scene. If faces are shadowed, it is when the entire figure is in shadow (as in some of the shots in the first escape scene).

Thirdly, all these darker scenes are scenes of danger. This is the unambiguous meaning of all the scenes and shots noted above. Thus obscurity and ambiguity, frequent connotations of low-key lighting, are totally dominated by the force of the primary connotation.

The most complex camera movement occurs at the openings of scenes. The first scene in Nottingham Castle, when Sir Guy is entertaining Prince John, begins with an 11-second high-angle pan following four servants across the hall. This is followed by a 36-second crane shot. It begins at a torch burning on a wall, moves down to the fireplace and then travels parallel with another group of servants across the hall. Leaving them, it continues along and behind a row of musicians, soldiers and finally guests. The shot ends by tilting down to show a dog gobbling meat on the floor (at the words 'Hail Prince John') before cutting, none too subtly, to Prince John eating. Although this opening is more elaborate than most, its combination of crane movement and partial following of characters is fairly typical. Some opening shots, however, are completely free of significant character movement. The final scene of the film begins with a 29-second crane shot which starts looking down at a pile of swords and shields and ends at a high angle, looking down on a motionless crowd. The only character movement in the shot is in the background and is unrelated to the movement of the camera. Such shots only occur as establishing shots, surveying a scene and providing impetus at the beginning of a sequence. They are spectacular in the most literal sense – they reveal the spectacle. They are minor, if noticeable, parts of the film, but meaning is contained and limited by their conventionalised use.

The most frequent departure from the more restrained style is the use of high- and low-angle shots but, here as well, their occurrence is in fact very limited in function. Just over a quarter of the shots in the film are angled in some way but less than 3 per cent are completely unmotivated by character position. Typical shots look up from normal eye-height to a figure on horseback (as in the first shots of Robin). High angles are conventional crowd scenes, with the camera raised to a height from which the extent of the crowd can be seen. Since, like most swashbucklers, the film is concerned with conflicts between figures of power and authority, the low angles mainly look up at these figures. There are no unmotivated extreme angles. The strongest angles are down from, or up to, the trees during the forest ambush and the exterior shots at the meeting of Robin and Marian in the castle mentioned earlier, in which the camera looks down at Robin climbing up the wall. The camera is nearly at an overhead position but is motivated by Marian's location as she looks out of the window.

Other features of note include several canted shots used to build up excitement (during the scene of Robin's rescue from the gallows and during the archery contest) and one shot with a look direct to camera. This is of Sir Guy and occurs in the final duel just before he is killed. It is very brief and is strongly locked into a reverse-cut structure, being preceded by a CU of Robin and followed by a two-shot in MS.

The Adventures of Robin Hood, then, shows how the restrained style of the later middle thirties could include some elements which went beyond the simple style of *The Awful Truth*, but which are, nevertheless, firmly anchored by the dominant style, tending more to emphasise it by showing its outer limits. Not only are these features overwhelmed by the dominating style, but the attempt is made to restrict their significance by reducing ambiguity to a minimum.

The Adventures of Robin Hood is in colour and this has two consequences which are of relevance in this context. The first is that the sense of spectacle is increased – quite intentionally in order to help recoup the extra cost entailed by the use of Technicolor. This may have given an added reason for the spectacular crane shots which show off the detail of the highly-coloured sets. The second consequence is that high-key lighting becomes even more likely in order to reveal the sets. The few darker scenes become all the more surprising in view of this, particularly since the main function of the Technicolor representatives on the set was to ensure the adequate registration of colour – in other words, to show off their product at its best. The fact that a few dark shots did reach the release print is perhaps due to the amount of colour in most of the other scenes, particularly the scenes of pageantry such as the archery contest. In addition, of course, at least one of the darker scenes (that of the sunset execution) features a particularly dramatic use of colour.

THREE COMRADES

Three Comrades, made by MGM, was premiered in June 1938, just one month after Warner Brothers released *The Adventures of Robin Hood*. Produced by Joseph Mankiewicz and directed by the romantic melodrama specialist Frank Borzage, it is based on the novel of the same name by Erich Maria Remarque. It is perhaps best known now as having a script which contains F. Scott Fitzgerald's most substantial contribution to any Hollywood film. The cinematography was by Joseph Ruttenberg (who received the 1938 cinematography Academy Award for his work on another MGM film, *The Great Waltz*). *Three Comrades* received one Academy Award nomination – for Margaret Sullavan as Best Actress.[9] Matthew J. Bruccoli notes that it 'was a marked success, making the best-10 lists for the year.'[10]

Three Comrades begins where *All Quiet on the Western Front* ends. The opening three-and-a-half minute scene takes place in November 1918. Three German soldiers – Erich (Robert Taylor), Gottfried (Robert Young) and Otto (Franchot Tone) – are having an end-of-war drink and are looking to the future. The rest of the film takes place in 1920 when the three friends are running a taxi/garage business. They meet Pat (Margaret Sullavan) and Erich falls in love with her, marrying her in the summer. The second part of the film is concerned with two tragic events – Gottfried's death at the hands of a political opponent (which is followed by Otto's avenging murder of the assassin) and the death of Pat in a sanatorium at Christmas. The film ends at Pat's death, with two of the four main characters dead and the two survivors returning to the political battle going on in the cities. It is thus precisely the kind of tragic melodrama which might be expected to use 'expressionist' elements such as low-key lighting, unusual camera angles and complex camera movements. In fact, there is little of this in *Three Comrades*. This can best be seen by considering the three most melodramatic moments – Gottfried's death, Otto's revenge killing and Pat's death.

Gottfried is killed during a riot at a political meeting. The sequence is as follows, beginning immediately after Otto and Erich bid farewell to Pat as she goes off to a sanatorium for the winter.

Shot 1 High angle of a crowd pushing at a door, trying to get in to a political meeting. Pan round to reveal Otto and Erich arriving at the scene. The camera then moves down and in to a MLS of the two friends, losing the high angle.

Shot 2 Exterior shot of the back door of the building in the previous shot. A slight low angle looks up from street level to the door, a few steps above. Some men come out in a hurry.

Shot 3 Extreme high angle looking down on the men coming out of the door. In the foreground is a first-storey balcony on which a man with a rifle appears. The escaping men disappear under the camera and the man on the balcony takes aim at them. Otto and Erich arrive in the street and see the man shoot.

Shot 4 LS of Gottfried, the last of the escaping men, being shot. The angle is parallel to the ground and corresponds to the point-of-view of Otto and Erich.

Shot 5 MS of the murderer, at a slight low angle. He sees Otto and Erich looking at him.

Shot 6 Reverse cut. MS high angle of Otto and Erich.

Shot 7 MS of the murderer, at a low angle. He goes back inside the house.

Shot 8 MLS of Otto and Erich, with the angle parallel to the ground. They run forward.

Shot 9 LS of Otto and Erich running up to Gottfried (set-up as in shot 4).

Shot 10 MCU three-shot of the friends. Gottfried dies.

Shot 11 MS of the three men with Dr Becker (Henry Hull), Gottfried's political mentor, standing behind.

Shot 12 MCU three-shot, as in shot 10.

It will be clear from this description that all the angles are motivated, either by the position of the man on the balcony or by the position of the men in the street. The only exception is the opening high angle but this is a conventional crowd view which moves down to a 'normal' angle as soon as the main characters appear. There are no ECUs in the scene, despite the emotional intensity of Gottfried's death. Most notably, every shot is in high-key lighting (and there is no reason why the scene could not have been set during the evening if motivation for dark effects had been desired). The only camera movement is in the opening shot and it is simply the conventional movement within an establishing shot, moving from a general view of a scene into a detail within that scene. Admittedly some darker scenes follow this, as Otto and Erich hunt for the killer, but the point is that the scene of high emotion is well-lit.

Not long after that is the scene in which Otto finds and kills Gottfried's murderer. This is one of the most dramatic sequences in the film. It begins with Otto waiting in his car outside a house in which he knows that the murderer is meeting with friends. The scene is set in snow at night.

Shot 1 Some men come out of a building and down a flight of exterior stairs. There is a quick pan round to reveal Otto in his car. He is in MS, with the camera at a slight high angle (from the height of someone standing beside the car). His face is in light although the rest of the frame is fairly dark.

Shot 2 LS of the five men at the bottom of the stairs. Four go one way, the fifth (the murderer) goes the other way.

Shot 3 MCU of Otto in his car. His face is well lit, but for a band of shadow across his mouth (caused by the car windscreen). The background is very dark.

Shot 4 The murderer in LS as he looks back hearing Otto start his car.

Shot 5 Otto in MS. His face is clearly lit but the background is in shadow. The camera pans round as the car moves forward.

Shot 6 Tracking shot from Otto's point of view as he follows the murderer (who is in LS). The car headlights throw the murderer's shadow on to the wall behind. As he becomes aware of his pursuer, he runs down an alley. Car (and camera) stop and Otto appears in the frame and goes down the alley, throwing a very large shadow on to the wall as he does so.

Shot 7 The murderer (in MS) stops and looks back. There is some light shadow on his face, a strong light in the background and a strong shadow beside him in the middle ground. He runs off. Otto appears, walking firmly after him.

Shot 8 The murderer in LS appears round a corner. He tries a locked door and then runs up to a large double door at a church. The camera tracks with him. The door is locked and he runs out of the frame. Otto walks determinedly across the frame.

Shot 9 The murderer in LS runs down a narrow alley to a dead end. He turns and shoots. Otto shoots back at him although only his shadow is seen. The murderer falls. The alley is lit in a fairly light shadow with some strong lights.

Shot 10 A low angle MLS of Otto, a few steps above the camera, looking down the alley. He turns and goes.

An important constituent of the drama of this scene is the soundtrack. To begin with, only some muted effects are heard (most notably the car engine). Then, at the end of shot 7, the final part of Handel's 'Hallelujah Chorus' is faded in, ending at the close of shot 10. This is by far the most dramatically staged scene in the film, yet even here camera movement is all motivated by character movement, the camera angles are all at least weakly motivated and the range of distances is from LS to MCU. The lighting stands out but it is not particularly surprising. The wall shadow in shot 6 has a clear source, and in no shot are the eyes of either character in shadow, except very briefly in shot 7 as the murderer is moving and in shot 9 when he is too far away to be clearly seen. The bar of shadow across Otto's face in shot 3 is his only strong facial shadow.

Pat's death occurs right at the end of the film followed only by a brief four-shot scene of Erich and Otto. It is preceded by the last scene of Erich and Pat together – a scene shot entirely in high-key lighting. After Erich leaves Pat's room, the following scene is shown.

Shot 1 Pat in bed in CU, the camera looking down at her.
Shot 2 MS higher angle of the same. The camera tilts up slightly as she gets up from the bed and then it moves to a near overhead position as she stands up. She walks slowly across to the window with the camera going to a slightly higher position and tilting after her. A shadow from the window frame is cast on the floor.
Shot 3 Exterior low-angle shot looking up to Pat emerging on to her balcony in LS.
Shot 4 Reverse cut looking down past her back (in MS) to Erich saying goodbye to Otto in the distance (in ELS). Otto leaves.
Shot 5 High-angle LS of Erich as he catches sight of Pat.
Shot 6 Low-angle MLS of Pat as she stretches out her hand.
Shot 7 Erich, as in shot 5.
Shot 8 Pat as in shot 6, but with both arms stretched out to him. She collapses.
Shot 9 Erich, as in shot 7. He runs towards the building.
Shot 10 Interior of the sanatorium's hall. Erich runs in and up the stairs, the camera panning round and tilting up with him.
Shot 11 Pat lying on the floor in MLS. Erich enters the room and kneels beside her.
Shot 12 MS two-shot.
Shot 13 CU of Pat, over Erich's shoulder, seen from a high angle.
Shot 14 Pat and Erich, as in shot 12.
Shot 15 Pat, as in shot 13. She dies.
Shot 16 MS as Erich embraces her.

Again music plays an important part, with quotations from the 'Liebestod' from Warner's opera *Tristan und Isolde* included in a score which begins quietly in shot 1 and ends as Pat collapses in shot 8. In this scene all shots are in high-key lighting and, as will be obvious from the description, all angles are motivated except for those in shot 2. This is an extraordinary shot, with the unusually high angle standing out from the rest of the scene (in fact, from the rest of the film), and yet the remainder of the scene is shot in a completely conventional style. Even allowing for the typical Borzage theme of love transcending death (and therefore making Pat's death in a sense less tragic), the style of this scene is remarkably restrained, and in this it is typical of the rest of the film. There are some darker scenes (for example, the race through the night to get a doctor to Pat after she has collapsed on her honeymoon) but these are brief and seldom extend to dark faces. The only other breaks in style are the montage sequences by Slavko Vorkapich, in which subject positions constructed by the rest of the film are disrupted but only in a highly conventionalised way. These montage sequences are clearly marked as different and as very self-conscious narrational devices. It is as though they are in inverted commas and their disruptive effect is thereby minimised.

It might be suggested that rather than being typical of the period, *Three Comrades* is only typical of its director, Frank Borzage, and so a comparison with an earlier Borzage film is instructive. *A Farewell to Arms* has much in common with *Three Comrades*. It is based on a literary source – Ernest Hemingway's novel. It is set against a background of important historical events (the end of the First World War). It concludes with the heroine (Helen Hayes) dying in the arms of the hero (Gary Cooper) and even uses music from *Tristan und Isolde* for the death scene. The differences are that it was made at Paramount with Charles Lang as cinematographer and, most importantly, it was made in 1932, being released in December of that year.

Lighting in *A Farewell to Arms* is often dark with a heavy use of shadow, especially on faces. The camera is frequently angled and completely unmotivated angles are not uncommon (for example, during an early scene of the hero, Frederic, and the heroine, Catherine, walking together at night, there is a high angle shot looking down on them through the legs of an equestrian statue). In addition to this, the film includes a remarkable subjective sequence as Frederic is wheeled into the Milan hospital where Catherine is working. The sequence includes three shots directly from his point of view and with other characters looking at him. The first is a long travelling shot looking up at the ceiling from the trolley on which Frederic is lying. The second, in Frederic's hospital room, ends with Catherine coming up to the camera to kiss Frederic and appearing in a very big ECU – one eye only. These two shots together last for 109 seconds. There is then an 18 second two-shot of the couple with the camera looking at them side-on, before the third subjective shot which continues the set-up of the second shot and lasts for 19 seconds. Although the whole scene is not particularly long, it is sufficiently striking to stand out clearly in the film as a whole.

The comparison should make it clear that the style of *Three Comrades* cannot be reduced to a sign of directorial control. *Three Comrades* is typical of the later middle thirties, just as *A Farewell to Arms* is typical of the early thirties. An expanded stylistic paradigm and a willingness to experiment are as typical of the early thirties as the restrained high-key style is of the 1936–38 period.

IDEOLOGY IN *THREE COMRADES* AND *THE ADVENTURES OF ROBIN HOOD*

Three principal ideological elements can be noted in *Three Comrades*. The first is the suppression of politics. This might at first seem an odd assertion to make. After all, the film takes place against a background of political turmoil and Gottfried dies because of his political beliefs. Despite this, it would not be easy for an uninformed spectator to identify the opposing groups. No political parties are mentioned, nor are any general descriptive terms used. There are hints in the text, but they would only be meaningful to a spectator who knew something about the political conflicts in Germany in the interwar period. Thus Breuer angers Gottfried by saying, 'What Germany needs is order, discipline.' Gottfried describes Pat as 'a rich man's girl'. From hints such as these it is possibly to identify Gottfried as left-wing and his opponents as right-wing, but this is never spelt out. To the unsophisticated mass audience in the United States, the politics of the film must have seemed impenetrable (and, indeed, irrelevant).

The second ideological element in *Three Comrades* concerns the treatment of women. At a fairly obvious level Pat is separated from the three friends. The title refers only to the men (and behind the opening credits is a picture of three male heads). Pat is put on a pedestal by all three and is a passive character, reacting to the men rather than initiating action herself. At a more insidious level, it is worth considering the role her illness plays. The first sign of her illness is when she coughs after having been waiting on the steps outside Erich's rooms. This waiting is the first unambiguous sign that she loves Erich. Her collapse occurs on their honeymoon as she is taking part in physical exertions with Erich. Both of these incidents suggest that at some level her illness is linked to her relationship with Erich. Mary Ann Doane has discussed such links in the context of Hollywood in the forties. 'A woman's illness may be defined in many ways by the classical text, but it is never simply illness. More often than not it is a magnification of an undesirable aspect of femininity or a repudiation of femininity altogether. If the disease is excessive, it invariably necessitates punishment – in the interests of a legible sexual differentiation.'[11] Pat's transgression is clear – at the beginning of the film she is Breuer's mistress (although, like the politics, this is not stated unambiguously). Not only that, but her marriage to Erich breaks up the trio of male friends. If the film is about the friendship of three men (as the title seems to indicate), then Pat is a destructive force. What this indicates is that the sexual bias of the film is very firmly against women, and at a fairly profound level in the text. Pat's illness becomes, in Doane's terms, a manifestation of her sexual threat which becomes manageable for the film only by her death which becomes the inevitable punishment, restoring male control.

The third ideological aspect worth considering is the stress on individualism as the ultimate value, perhaps expressed most clearly in Pat's comment on Gottfried. – 'He did what he thought was right.' Each of the three main male characters is defined by his individual aim – Erich's love for Pat, Otto's love for his car and Gottfried's involvement in politics (Otto and Erich's lack of interest in politics serves to isolate it as a particular interest of one person, rather than being anything more pervasive). The lack of information about the characters' social backgrounds serves to emphasise individualism by isolating them from any kind of social determination. Only the war functions as a determinant, yet even it is, in effect, devalued. Otto comments about the cynicism of himself and Gottfried, having been in the war longer than Erich, and yet Gottfried dies for his beliefs and Otto is prepared to sell his beloved car in order to raise money for Pat's operation. The overall message of the film (underlined by the final shot with the ghostly appearances of Gottfried and Pat) is that the will of the true individual can conquer all – even death. Thus it conforms to the narrative of celebration, even though it has what seems to be a tragic ending.

Much the same selection of themes appears in *The Adventures of Robin Hood.* That politics is suppressed in a film set in medieval England may seem unremarkable. However political issues are quite overtly raised. This happens most obviously in the scene in which Robin shows Marian the poor, made homeless by the Normans, who follow his band and whom he feeds, but the references to heavy taxation, corrupt government and a leader who is more concerned with overseas wars, would each have a contemporary resonance for the late thirties American audience. Despite this raising of political issues, the suppression which was evident in *Three Comrades* happens here as well and is made all the more effective by the issues being raised in the first place. The mechanism by which the suppression is made is essentially the same one as in the MGM film – political problems are translated into personal ones. The problems of twelfth-century England are portrayed as being essentially caused by King Richard's prolonged absence abroad and the greed of Prince John and his accomplices. This does not destroy the film's relevance for Depression America but rather makes it all the more interesting. Ina Rae Hark has argued that *The Adventures of Robin Hood* is part of the Warner Brothers tradition of socially-conscious films, based as it is on 'the efforts of a charismatic individual to restore responsible government and economic stability to his country'.[12] In addition, the depiction of life under a brutal dictatorship is related to events in Europe. Whichever aspect is considered, however, the real hero is seen as Roosevelt. 'Whether *Robin Hood* represents a retrospective look at America's emergence from economic disorder or a prophetic glance at World War II, the heroes Robin and Richard in large part stand in for Franklin Roosevelt.'[13] This reading links up with the theme of individualism, which is taken a step further than in *Three Comrades* by the emphasis on charismatic leadership and the final submission to due authority. What Hark fails to note is that by translating political issues into problems of personal characteristics, an ideological move is made which pushes the film away from any consideration of social, economic or even political issues. All problems ultimately become personal ones and so the vagaries of human nature are the only relevant causal factors.

The treatment of women in *The Adventures of Robin Hood* is fairly straightforward. Maid Marian is treated in a similar way to Pat, being put on a pedestal and being essentially a passive character, reacting to the menfolk rather than acting on her own behalf. Her role is more clearly seen when Robin shows her his dispossessed followers being fed. She becomes the representative of those who need to be led, need to be shown the truth by the more aware male leader. The main difference between the roles of Pat and Marian is that the latter does not pose any threat to the men. Despite Robin's dangerous attempts to see her in the castle, the only time where he is likely to desert his men is at the very end of the film when the danger is over and Richard is restored. Only then does Robin disappear with Marian. It is not surprising then that Marian lacks the more extreme signifiers of illness and death which mark Pat.

The relationship between content and style should by now be clear. The signifying codes at work in these two films emphasise openness and clarity through the high-key lighting. There is no suggestion of anything hidden or obscure (unlike, for example, the indirect hints about Helga's upbringing in *Susan Lenox*) or contradictory (such as the conflicts between morality and the law in *Les Miserables*). Camera movement and camera distance underline the importance of the individual, with the camera only departing from characters in highly conventionalised situations (as in establishing shots). Camera movement is so restrained that it comes to signify stability, alongside its emphasis on character movement. Camera distance avoids extremes of closeness, allowing an emphasis on the active individual. Unmotivated angles, like camera movements, only appear in conventionalised settings (such as high-angle shots of a crowd scene). Predictable structuring and unproblematic subject construction round off the account of this style.

When mentioning the unproblematic subjectivity, it is worth emphasising that the implication of this is not that it only allows one reading of the film, or imposes values on to viewers, but rather that it makes non-preferred readings more difficult to achieve – at least for the original audience. In addition, it does not question the viewer's relationship to the text by putting him or her in diegetic subject positions (through the use of direct looks to camera). The viewer is always left in the safe position of the voyeur – looking, but never looked at. Thus not only does the style not hinder the ideological content, but it actively promotes it.

CONCLUSION

It remains to relate these findings to the earlier discussion of ideology in the mid-thirties. There it was concluded that this period was one of reassertion, although this did not indicate a single, unified response. Hollywood, on the other hand, was marked by a retrenchment. The reassertion became a very conservative interpretation of what traditional values were. Whereas in the country at large disagreements continued under the domination of the official White House creed of optimistic, liberal-democratic capitalism, in Hollywood the attempt to appeal to the greatest audience was interpreted as the need not to offend anyone and this resulted in an ideological cohesion which contrasted with the state of the nation. This cohesion in the content of Hollywood films in the mid-thirties has been noted by others. Andrew Bergman has talked of the 'evasions of the films of the mid thirties', films which 'mainly streamlined and reinforced the old goals'.[14] Nick Roddick, talking specifically about Warner Brothers films of the period, notes how 'individual motivation becomes the keynote' and 'society becomes the background rather than the conditioning factor'.[15] Garth Jowett comments that the only problems the Production Code Administration had to deal with in 1937 'concerned the depiction of foreigners and the "excessive drinking" in American films'.[16] What this chapter and the previous one demonstrate is that a restrained cinematographic style also had a role to play, adding to the cohesion of the content. The restrictions of the later mid-thirties covered not only *what* was seen, but *how* it was seen.

9 Towards *Film Noir*: *Dead End*

At the end of the thirties another set of changes began to take place in Hollywood cinematographic style. By 1939 they had become fairly widespread but their origins can be found earlier, even as the restrained style was in full bloom. *Dead End*, made at Samuel Goldwyn's studio and directed by William Wyler, will illustrate this development. Although made before 1939, it shows clearly the way in which the changes would move and also serves to warn against any too rigid division of the decade. Ideological changes did not take place suddenly, affecting all films at the same time. Rather the differing industrial contexts in which films were made allowed differing rates of reaction.

THE MAKING OF *DEAD END*

During the 1930s United Artists faced a continuing problem.[1] The company needed more films to distribute. It had been formed in 1919 by Douglas Fairbanks, Mary Pickford, Charles Chaplin and D. W. Griffith with the intention that each of these would produce his or her own films. Griffith's career was fading during the 1920s and after 1930 he completed only one film for the company (*The Struggle*, 1932). Fairbanks and Pickford both suffered from the transition to sound and changing public taste. Fairbanks' last United Artists release was *Mr Robinson Crusoe* (1932), a mixture of travelogue, Fairbanksian athleticism and very little story. Mary Pickford attempted to change her screen image from her little girl persona but was unsuccessful. She only appeared in one film during the thirties (*Secrets*, 1933) although she also co-produced two films with Jesse Lasky in 1936. Chaplin could still produce successful films but his work-rate was extremely slow, resulting in just two films during this decade – *City Lights* (1931) and *Modern Times* (1936). Although the former was a great success, the latter only covered its costs after distribution abroad. Thus the problem for United Artists was to keep up a regular supply of product when the company's founding stars were inactive. To do this the company added other independent producers to its membership.

The most important of these later additions was Samuel Goldwyn. Goldwyn joined United Artists in 1925 and during the 1930s he was releasing three to five films each year (apart from 1930 when he released just two and 1931 when he released six). Although Twentieth Century, the Joseph Schenck/Darryl F. Zanuck company, released 18 films through United Artists in the years before its merger with Fox (1933–35), by 1936 Goldwyn was the leading producer with United Artists and he retained this position throughout the rest of the decade, being challenged only by David O. Selznick (from 1935) and Walter Wanger (from 1936). He was constantly at odds with Chaplin, Fairbanks and Pickford, producing the films which kept the company in the public eye, while the others were taking their share of the profits. This situation led to his unsuccessful attempt in May 1937 to buy the others out.

The position in which Goldwyn found himself put a certain amount of pressure on him. He needed successful pictures to bolster his economic argument against his partners. However, he also needed prestige films in order to compete with people like Chaplin and Zanuck. Indeed in the 1930s he had to do this simply to justify his position as an independent but non-Poverty Row producer. He had to differentiate his films from those of the large studios and this could only be done in terms of quality if he was to justify his involvement in United Artists. It was out of this situation that the contradictions of Goldwyn emerged – he kept a close eye on all aspects of his productions but was prepared to let directors spend money in a way that few other producers would allow;[2] his ignorance and philistinism were legendary and yet he produced some of the most daring films (both in content and in style) of the second half of the thirties

As head of a small studio Goldwyn was able to build up something of a family atmosphere. He developed a team of top professionals who were encouraged to produce innovative work. The team included director William Wyler, cinematographer Gregg Toland, composer Alfred Newman, designer Richard Day and editor Daniel Mandell. This group was responsible for *These Three* (1936), *Dead End* (1937) and *Wuthering Heights* (1939). The same team but with Rudloph Maté replacing Gregg Toland made *Dodsworth* (1936). James Basevi (special effects designer on *Dead End*) replaced Richard Day on *The Westerner* (1940) and he in turn was replaced by Stephen Goosson (already an Oscar winner for his work on Columbia's *Lost Horizon* in 1937) on *The Little Foxes* in 1941. (This last film was released by RKO, coming after Goldwyn's final break with United Artists in 1941.) In addition to those already mentioned, the name of Lillian Hellman, scriptwriter on *These Three*, *Dead End* and *The Little Foxes*, should be added.

Most of these films were based on literary sources. Sidney Kingsley's play *Dead End* had been a Broadway success in 1935 and on buying the rights (for an exceptionally large sum)[3] Goldwyn agreed with Wyler and Hellman to make a faithful adaptation.[4] Some changes, however, had to be made to get the script past the Hollywood censors. The Production Code would not have allowed the play's freedom of language and many references needed to be toned down (for example the fact that Francey has syphilis is veiled beyond recognition in the film). Hellman changed some of the characters, with the play's central figure of Gimpty, a bitter cripple, being transformed into a more conventional leading man.[5] However, the structuring oppositions of the play were retained – the juxtaposition of the very rich and the very poor, the kids and the adults, the adults who were successes and those who were failures.

Wyler originally wanted to shoot the film on New York locations – an idea whose originality it is now difficult to appreciate – but Goldwyn vetoed this, apparently for fear of losing control of the production.[6] A large set designed by Richard Day was built in the studio, becoming, according to Madsen, 'the talk of the town'.[7] (The opening shot of the film is no doubt motivated as much by the desire to show off the set as it is by narrative purposes.) Gregg Toland contributed much to the film as well. Anderegg quotes Wyler as inscribing Toland's copy of the script with 'your co-director, Willy Wyler'.[8]

The cast included the kids from the Broadway production – a decision which, more than any other single factor, gave the film what non-Hollywood quality it had. The boys went on to make *Angels With Dirty Faces* for Warner Brothers in 1938 and then to a number of lesser films, such as *The Angels Wash Their Faces*, but in *Dead End* they were new and presented a screen image very different from the prevailing Hollywood norms as seen in such child stars as Mickey Rooney, Freddie Bartholomew and Jackie Cooper. The two leading adult stars – Sylvia Sidney and Joel McCrea – were Goldwyn contract players at the time.[9] For the role of the gangster, Goldwyn brought in Humphrey Bogart from Warner Brothers. In the late thirties Bogart was typecast as a vicious gangster, largely due to his 1936 role as Duke Mantee in *The Petrified Forest*.

Dead End was premiered in August 1937 and was an immediate critical success, getting Academy Awards for Best Picture, Gregg Toland's cinematography and for Claire Trevor, in her four-and-a-half minute role as Francey, as Best Supporting Actress. It was a reasonable financial success although not featuring in the top fifteen money-earners of 1937–38.[10] Madsen notes that according to Georges Sadoul, this was the film which originated Wyler's reputation in France.[11]

THE SCRIPT

The story of *Dead End* can be broken down into three narrative strands. The fluidity of scenic construction makes this an easier way of grasping the film's structure than a straightforward line-by-line account. The setting is a dead-end street in Manhattan's East Side, looking out over the East River to the Brooklyn Bridge. The action takes place within one day.

A gang of five kids live in the street – T. B. (played by Gabriel Dell), Spit (Leo Gorcy), Dippy (Huntz Hall), Angel (Bobby Jordan) and their leader Tommy (Billy Halop). A new kid – Milty (Bernard Punsley) – arrives in the street and is allowed to join the gang. The kids beat up and steal a watch from a rich boy who lives nearby (Charles Peck), but his father, Mr Griswald (Minor Watson) catches Tommy, who escapes by stabbing Griswald's hand with a penknife. Spit is caught by the police and informs on Tommy but, before the police can catch him, Tommy has trapped Spit and nearly injures him. Tommy is finally led off, facing the prospect of reform school.

Intertwined with this is the story of Baby Face Martin (Humphrey Bogart). Now a famous gangster, Martin returns to the street of his youth. He is accompanied by his henchman, Hunk (Allen Jenkins). He first meets a boyhood friend, Dave (Joel McCrea) who is an unemployed architect, and then meets his mother (Marjorie Main) but is rejected by her. Next he finds his childhood sweetheart Francey (Claire Trevor) but rejects her when he realises that she is now a prostitute. In return for these disappointments he decides to kidnap the rich boy but Dave interrupts this plan and shoots Martin. Martin dies and Hunk is led off by the police.

In addition to these two strands there is a romantic plot. Dave and Drina (Sylvia Sidney) are childhood friends. Drina (Tommy's elder sister) loves Dave but he has fallen for Kay (Wendy Barrie) who has escaped from poverty by becoming a rich man's mistress. Dave finally realises that Kay is not for him and as the film closes he takes Drina's hand, offering the money he is due as a reward for killing Martin as a means of helping Tommy.

It will be clear from this outline that Dave is the central character, particularly since he intervenes in the first plot at a crucial point (stopping Tommy from marking Spit's face with his knife) as well as being involved in the other two. Dave functions as the moral centre, the conscience of the film, trying to educate the kids, killing Martin, and finally preferring Drina to Kay. Drina is also important although given a more emotional and passive role.

Much of the social criticism in the film is left-wing, although of a reformist rather than a revolutionary character. Poor housing is seen as the cause of crime. Reform schools are seen as schools of crime. Drina is a striker on picket duty and makes an impassioned defence of striking, also attacking policy brutality (her role here is without equal in thirties Hollywood). There is even an implied critique of male double standards. When Martin realises that Francey is a prostitute he recoils in disgust.

Martin: Why didn't you get a job?
Francey: They don't grow on trees.
Martin: Why didn't you starve first?
Francey: Why didn't you?

There are also, however, other elements which suggest a different reading. Mr Griswald is given a sympathetic appearance in his final speech, defending the need to prosecute Tommy. Martin is made an extremely vicious character, suggesting that it was not just bad housing that sent him wrong but that he is a 'natural' psychopath. And of course the very existence of Dave – a slum product who did *not* go wrong – undermines the force of the film's apparent argument. As already noted, Dave is the moral centre of the film and this dissipates the effect of Drina's speeches. He is never heard defending strikers. The film has a near happy ending as Dave and Drina smile happily at each other and there is at least the possibility of Dave's reward money helping to keep Tommy out of reform school. Finally it is worth noting that no general, structural causes are suggested to explain slum housing – it is merely there as a fact of life. The criticism of the rich seems to be only that they are insensitive in displaying their wealth near the poor. There is no hint at any kind of radical redistribution of wealth.

Thus it is a film with genuine radical leanings but with a compromised stance. It is instructive to compare it with *Angels With Dirty Faces*, made a year later at Warner Brothers and directed by Michael Curtiz. All five of the 'Dead End Kids' appear playing essentially the same characters as in *Dead End* but given different names. Billy Halop is still the leader of the gang but now he is called Soapy. Leo Gorcey is still the most rebellious and aggressive gang member but is now called Bim. The later film starts with a crane shot across a slum street in New York. It is an open-air set and looks much more realistic than that of *Dead End*. This might lead one to expect the later film to be more radical and uncompromising than the earlier one but this is not the case. *Angels With Dirty Faces* is an unambiguous film. Crime of any kind is wrong and no social causes are suggested. The central opposition between Rocky Sullivan (James Cagney) who is a confirmed gangster and Jerry Connolly (Pat O'Brien) who is a priest, functions as a way of denying the effect of the environment since the priest goes through the same deprived childhood as the gangster. The final transformation of Rocky when he pretends to be scared of the electric chair so as to prevent any hero-worship among the boys, is the triumph of a simplistic black-and-white morality. The assumption of *Angels With Dirty Faces* is that there is no excuse for crime and that the solutions – religion and the law – are unquestionable. *Dead End*, at the very least, raises unanswered questions about such a simplistic account.

Dead End, then, was a film which Goldwyn wanted to be both a financial and a critical success. It was made by a team who had already made quality films together (including at least one – *These Three* – on a controversial subject) and who were encouraged to innovate. The film contains internal contradictions – it is one of the most powerful left-wing social critiques to emerge from Hollywood in the thirties and yet it allows plenty of scope for a rejection of that critique. It therefore provides a particularly interesting example of the interaction of style and content.

CAMERA MOVEMENT

The statistics for camera movement over two 100-shot sequences are as follows.

Static shots	59.5%
Reframing shots	13%
Pans and tilts	22%
Tracking and crane shots	5.5%

This is a fairly low figure for wholly static shots but is balanced by a low figure for travelling shots. This latter figure may seem surprising since the general impression of *Dead End* is of a film with a fairly mobile camera. The reason for this impression is that the travelling shots are very noticeable. They are frequently long takes, unlike the moving shots in *The Awful Truth*. Their principal function is to link scenes and thus move the narrative on smoothly.

A typical example of this technique links the scene of Martin and Hunk in a restaurant. The shot begins by showing Martin, viewed from directly across the street, coming out of the door of the tenement in which his mother lives. As he walks towards the camera, it moves back and is revealed to be behind a window. Continuing back, it picks up Martin and Hunk as they come through the door of the restaurant and sit down at the counter. The camera then pans round with Hunk as he goes over to the pianola and then pans back with him as he returns to the counter. Finally it pans back again to the pianola as he walks back to it. At this point there is a barely noticeable cut. The shot lasts for 36 seconds with a further 11 seconds on Hunk after the cut.

A slightly more conventional shot, partly a standard parallel track and partly a scene-linking movement, occurs earlier in the film as the kids make fun of a rich old man. The camera tracks along with the old man as he walks along the street, assisted by his chaffeur and nurse. It then leaves the older trio and pans round to the kids as they mockingly hobble after the adults, ending with a short tracking movement alongside them. The shot lasts for 26 seconds. It ends with the kids meeting Milty, the new boy on the street, and thus effectively links the previous sequence with the one about to start.

The most obvious travelling shots in the film are the opening and closing shots. The opening shot (or rather, what appears to be the opening shot) includes a disguised dissolve and so is actually two shots but since the effect is of a single shot, it will be described here as if it were one. It begins with a rooftop view across the New York skyline. Titles reveal that the location is beside the East River and that the rich are living here, directly overlooking the poor. The camera then moves down vertically. Between moving below the roof level and arriving at a window there is a hidden dissolve at a point at which only a darkly lit brick wall is seen. This conceals the change from a model shot of the skyline to the life-size studio set. On reaching a window the camera moves sideways over the roofs of smaller buildings to a high-angle view of the street showing a policeman, a milkman and a servant cleaning a back door. There is then a cut to a parallel-angle view of the same two people. The total length of the sequence is 82 seconds, the titles taking up the first 28 seconds and the movement after the hidden dissolve taking up the last 40 seconds.

The film ends with a shorter version of the same movement in reverse. The kids are seen in ELS, the camera craning with them as they walk across the set. They move out of sight between two buildings and the camera moves up a side wall. The hidden dissolve again makes the transition to the skyline view. The time taken is 37 seconds with the dissolve coming after 25 seconds. Such timings can be compared with the film's average shot-length of about 11 seconds (Barry Salt gives a figure of 9 seconds as the average shot-length for this period as a whole).[12]

A less obvious but more radical movement (in terms of stylistic norms) is the unmotivated movement in the second scene of Hunk and Martin in the restaurant. Starting with an over-the-shoulder view looking at Martin from behind Hunk, the camera moves towards Martin and then round until it is looking at both men from a 90° angle. The shot lasts for 31 seconds, with the movement occupying the first 20 seconds. This movement is totally unmotivated and since it is in the middle of a scene it cannot serve the linking function mentioned above. It is used here for emphasis – this is the point at which Martin decides to kidnap the rich boy, thereby setting in motion the events which will lead to his death.

The statistics quoted earlier show a high percentage of pans and tilts (22 per cent). This is one of the factors which leads to the film giving the impression of having much camera movement. It is partly the result of the restricted set. The camera pans rounds this set, following people across it. Even the linking function of the tracking shots is more frequently performed by pans. In effect the pans are compensating for the limited setting, trying to create as much variation as possible within the confines of one street. Very few of the pans do not follow a character or a character's gaze.

CAMERA DISTANCE

The percentages over the two sample 100-shot sequences are as follows.

ECU	0%
CU	3.5%
MCU	12%
MS	19.5%
MLS	49.5%
LS	12%
ELS	3.5%

As in *The Awful Truth*, distances are dominated by the MLS and the MS, these accounting for almost 70 per cent of the shots. The few ELSs are views of the whole set, usually introducing scenes. The closer shots – CUs and, outside of the sample sequences, ECUs – occur in the dramatic moments in the second half of the film, beginning with the confrontation between Martin and Francey.

The fluid nature of the scenic construction means that it is even more artificial than normal to detach specific scenes for analysis. Despite this, the conventional structuring uses of distance can be recognised, with more distant shots for linking scenes, and closer shots for climaxes and emotional reactions. The scenes do not tend to be strongly structured, however, and the frequent scenes which feature a group of people means that the MLS/MS range is preferred.

CAMERA ANGLE

At a first viewing of *Dead End*, the most obvious stylistic feature will most probably be the frequent use of strongly angled shots. The statistics for the two 100-shot sequences are as follows.

Shots parallel to the ground	39%
Low-angle shots	40.5%
High-angle shots	20.5%

This shows the highly unusual fact that there are very slightly more low-angle shots than parallel-to-ground shots. Even allowing for an untypical selection of shots, there is clearly an exceptionally high number of low-angle shots in this film. The figure for high-angle shots is much smaller, but is still comparatively high. Of the low angles, 25 per cent are unmotivated, and of the high angles 5 per cent are unmotivated.

When the use of these angled shots is examined, two points becomes apparent. Firstly, there are very few low-angle shots of the kids but many low shots of the adults. Typically, parallel-to-ground shots of the gang are contrasted with low-angle shots of the doorman at the luxury flats, or of Martin and Hunk, or of Drina and Dave. In this way the angles frequently serve to ally the viewer with the kids, positioned as their equals, looking up to the various manifestations of adult authority. This opposition is extended to a more general one between the inhabitants of the street and their overlooking neighbours, most obvious in the conversation between the rich boy looking down from his balcony and the poor kids looking up from the dockside. Both views use very steep angles. There are no side views of both elements of the opposition, just the polarisation of extreme camera angles.

The second point to note is the use of some very extreme angles which are either unmotivated or only weakly motivated. One example shows Martin and Hunk looking up at a tenement. The camera is placed at foot-level, looking up at the two men in LS, with the tenement behind them. The composition is framed by an archway. The camera position is motivated weakly by the fact that Martin and Hunk are looking up at the tenement but the motivation is only weak because their view, and their viewing of it, could be shown without resorting to such an extreme angle. This extremity of the angle calls attention to the placing of the camera. To put it another way, the enunciation of the film is foregrounded.

Another example occurs later in the film and again shows Martin and Hunk in front of a building. This time the building is the one with the rich apartments from which the gangsters plan to kidnap the boy. A low-angle shot looks up at the two men, with the building behind them. There is a cut to much the same view but from a much more extreme low angle, shot from foot-level. The effect is almost that of the jump cut since the transition between the two shots breaks the Hollywood convention of significantly changing the horizontal angle between shots. Again the enunciation is brought forward. The film becomes the film-maker's discourse by the highlighting of technique.

High angles are also used in this way. There are two set-ups which are virtually overhead shots. As Dave and Kay walk to the riverside for their final conversation, there is a high shot looking down over the edge of one of the rich apartment balconies, with Dave and Kay walking below. The shot is totally unmotivated. A later shot (which recurs) is similar but includes weak motivation. The camera looks down on Tommy as he lies on the top of a building looking down on the other kids (as he tries to find out who gave him away to the police). As in the previous example, the strong angular lines of the composition add to the overall effect, creating a pattern of near-abstract effect.

Again the enunciation is put into the foreground. Just as the low-angle shots mentioned above can be explained by showing how characters are dominated by their environment, so these high-angle shots can be explained as showing the vulnerability and helplessness of the characters. However, this explanation is not enough to explain the extreme nature of these shots and their stylisation of design.

LIGHTING

> His style of photography would vary, just like my style of directing. In *Dead End*, we had a different style of photography than in *Wuthering Heights* or in *These Three*. Here, we were dealing with little girls' things. What was good was rather simple, attractive photography. In *Dead End*, we had flat, hard lights. We used open sun-arcs from behind the camera. We didn't try to make anybody look pretty.

This is William Wyler, talking about Gregg Toland.[13] In fact, several styles of lighting are used in *Dead End*. These merge into each other but five stages along the continuum can be identified.

In some scenes a very harsh, bright form of lighting is used. The most extreme examples are in the two scenes in which Martin meets people from his past. The scene in which he meets his mother is not long (it lasts for only 2 minutes 20 seconds) but is very striking. His mother is lit so that her face becomes an almost featureless mask; the harshness (from Martin's point of view) of her rejection of him is reflected in the lighting. Martin can see nothing of comfort in her face. In some of the shots in the scene, shadow in her eyes makes dark holes of her eye sockets. The other scene with this kind of lighting is the scene in which Martin meets Francey. Halfway through this four-and-a-half minute scene, Francey steps sideways into the sunlight and tells Martin to take a proper look at her. The lighting is suddenly harsh and unflattering. The difference is as if Francey is seen at first as Martin wants to see her and then, suddenly, as she really is. Again the face becomes almost a mask with the strength of the lights leading to near over-exposure.

The second style, and the one which is used most commonly in the first half of the film, is a form of bright, high-key lighting but with less modelling effects than might be expected. The large arcs which Wyler referred to result in a stronger but less focused type of lighting than conventional high-key. The traditional aims of high-key lighting – showing the whole set and separating the cast from the background – are observed. An interesting deviation from standard forms of lighting is the fairly strict observance of the direction of the light. The passage of the sun across the sky can be noted from the few light shadows which are always in the appropriate direction. This, along with the intensity of the light, gives a naturalistic effect, unusual in a film of this period.

The third style is glamour portrait photography ('star' photography). This is not particularly obvious but in many MSs and MCUs there is a move away from the lighting styles used elsewhere in the film, towards a conventional portrait style. The effect is strongest on Kay, combined as it is with a softer focus. On Drina it is fairly subdued but there is enough top light to add a slight 'halo' effect. On Dave this style is even less obvious but it is still present, with the use of top lights and modelling effects.

Much of the second half of the film uses a fourth style – a fairly conventional form of evening lighting, with faces brightly lit but with a certain amount of light-to-medium shadow appearing in the background. The contrast ratio is not by any means extreme but is noticeably greater than in the high-key lighting in the first half of the film. As with the earlier lighting, there is a certain amount of naturalism implied here, since the scenes lit in this way occur in the evening.

The final stage is classic *film noir* lighting, appearing here seven years before the conventional date for the start of that style and four years before its conventionally accepted predecessors (*Citizen Kane* and *The Maltese Falcon*). In the following chapter the historical aspect of this will be discussed in more detail and it will be enough at present to note the pivotal position of *Dead End*, coming after the classic gangster, horror and detective films of the early thirties and before the early examples of *film noir*. This style appears most strongly in the final third of the film. The scenes of the fight between Martin and Dave include many of the classic *film noir* elements – deep and impenetrable shadows, off-centre and unbalanced compositions, use of bars of shadow on faces, strong spotlight-type key lights.

Four shots will serve as examples of this style.

1 Kay visits the tenement in which Dave lives and is revolted by the filth and poverty. Dave watches her from a doorway. He is in MCU, just to the right of the centre of the frame. Half of his face is in shadow. There is a light

in the centre background. The rest of the frame is in darkness or near darkness.

2 Martin and Hunk are in an alleyway, just before Dave sees them and initiates their final fight. The two gangsters are in MLS. The left half of the frame is totally dark but there is some mottled light in its right side and a stronger light in the centre background. Martin and Hunk walk across the frame diagonally. Some weak light falls on their faces but mostly they are in darkness. Their eyes are never seen.

3 Dave is chasing Martin through a building. The only light is strongly directional and comes through two windows at the back of the set, creating silhouettes. Martin's shadow crosses the room.

4 Tommy, hunted by the police, is hiding in a tenement stair landing. The right side of the frame is in darkness. In the left side can be seen Tommy's face in CU, crushed by two bars of shadow.

As will be clear from these examples, this style is reserved for the most dramatic scenes.

The overall lighting pattern can be summarised as follows. The first part of the film, taking place in daytime, is in high-key lighting. The second part, taking place in the evening, is in a shadowy style but still preserves well-lit faces and enough light to see most of the backgrounds. Dramatic scenes in the first half use a very bright and very harsh form of facial lighting, while the dramatic scenes in the second half use the *film noir* style of low-key lighting. Within this overall plan, a form of weakened glamour lighting is used for the main 'good' characters, especially the women.

DEEP FOCUS

The most famous aspect of the style of cinematography associated with Wyler and Toland is the use of deep focus. As is well known, the classic form of extreme deep-focus cinematography only became a practical possibility in 1939 with the availability of faster film stock and coated lenses. *Dead End*, however, reveals a tendency towards deep focus even before the 1939 developments. Many shots in the film have two distinct planes of action. The nearest plane is not as close as is found in later films such as *Citizen Kane* and *The Little Foxes*. More typical of *Dead End* is the shot already noted of Tommy looking down on the other kids. Another example is when Kay visits Dave's tenement, in which the camera looks down on Kay and past her to Dave standing at the bottom of the stairs. Kay is not quite in sharp focus (although very near to it) but Dave is. Other shots have two planes of action but with one plane more obviously out of focus. When Martin and Hunk talk in the restaurant we see unfocused movement between them in the street behind. Many of the shots of the kids are composed in this way, with other people moving around in the background. It should be clear from this that the development in 1939 of the technical conditions which permitted deep focus cinematography was answering a need, rather than being a direct cause of stylistic change.

SIGNIFICATION

As has been already noted, the lighting sub-code used in *Dead End* is fairly complex. It is, however, subordinated to an overall progression from light to dark. As the plot darkens, so does the lighting. The darkest scenes are the narrative's most extreme and dangerous conflicts. The final scene is slightly lighter than the darker scenes preceding it. This pattern establishes the governing significations of the lighting. The traditional meanings of lightness and darkness are employed. Normality is light, darkness signifies danger, trouble, conflict and mystery. Even the two scenes with harshly bright facial lighting (Martin's meetings with his mother and with Francey) have darker backgrounds than the other settings in this part of the film. This systematic use of light and dark is, of course, extremely common in films (and in Western literature – 'the dark night of the soul') and therefore carries with it a traditional connotation. The narrative is fitted into a pattern of stability–conflict–resolution, and this pattern is reflected in the lighting. In this way an ideological pressure works against the more progressive implications of the film by fitting it into a conventional mould of the hero struggling with evil. The lighting follows the pattern of heroic endeavour and thus works to set the film in the style of a conventional narrative instead of allowing its inherent social criticism to come to the foreground.

There are, however, other aspects of the lighting which complicate this system. The dark *film noir* style which appears in the later parts of the film inevitably introduces ambiguity. The very fact that part of the frame is hidden introduces the unknown in a very direct sense. These dark areas of the screen can be seen as a mirror which allows an infinite number of meanings to be read into it (functioning a bit like the famous expressionless shot of Greta Garbo at the end of *Queen Christina*). How this works can be seen by considering the close-up of Tommy hiding, as described above. Part of the meaning of this shot is clear. Tommy is literally hidden and he is already imprisoned by bars of shadow. The shadow also isolates him, with the off-centre framing pushing him into insignificance. The ambiguity comes when the shot is read as containing an implicit judgement of Tommy. Is he basically a good character in trouble or a bad character in hiding? The narrative as a whole tends to the former but allows the latter. The ambiguity of the shadow permits (and to some extent encourages) both readings. The shadow can represent evil (or even just the environment) which crowds in on Tommy or it can represent the evil within Tommy as he becomes part of the shadow. In such ways does ambiguity enter into the text in the shadows of *film noir* lighting.

Another important aspect of the lighting is the attempt at naturalism, especially in the first half of the film. This may not seem very effective today but in the context of Hollywood in 1937 it represents a major departure for a feature film – a significant step towards the *plein air* lighting seen in documentaries such as *The Plow That Broke the Plains* (1936). It carries complex signification – realism, seriousness, social significance – all carried over from documentary and newsreel styles and all of which act against the overall light/dark system to emphasise the social implications of the text and play down its more conventional aspects.

Thus the lighting sub-codes of the film produce a complex effect. The conflict among the three systems of conventional light/dark patterning, ambiguous low-key and naturalistic lighting stands as witness to the conflicting ideological pressures at work in *Dead End*.

Camera distance functions to keep the viewer with the group, rather than the individual. There are few CUs, which might have pushed towards a greater emotional identification, and few ELSs which might have distanced the viewer from the events of the plot. Instead he or she is kept at a middle distance, always aware of the group of characters and the interplay among them. This helps to keep the social implications of the plot in the foreground, particularly since the group-setting frequently opposes the kids to the adults. On the other hand, attention is directed to the characters rather than the settings. This is of major importance in a film in which, ostensibly at least, the environment is the real villain. Thus the sub-code of camera distance can be seen as performing a balancing act, keeping away from revealing the setting too often or in too much detail. This balancing act reflects the overall ideological pattern of the film – social criticism balanced against Hollywood conventions.

As already noted, camera-angle represents one of the most radical aspects of the style of *Dead End*. It serves to link the viewer to the kids, with the adults being seen as powerful (and usually threatening) figures of authority. In addition it makes the environment appear threatening, with the low-angle shots frequently making the buildings loom over the characters. Instead of the setting being a neutral background (as in *The Awful Truth*) it becomes a participant in the drama. In this way camera-angle powerfully works to emphasise the critical aspects of the text. Even Dave and (to a lesser extent) Drina are distanced from the viewer by the angled shots.

The more extreme angles (from overhead and underfoot positions) function in a slightly different way. Because they are so extreme, they tend to remove the viewer momentarily from any identifications within the text. They do, however, tend to lead to identification with a fictional character *outside* the text – the film's author ('fictional' because the author is assumed simply through the mechanisms of the enunciation, not through any knowledge of the film's origins). It was noted earlier how these shots draw attention to the film's enunciation. In a 'social problem' film such as *Dead End*, this will tend to lead to an awareness of the status of the film as 'message' – someone is talking to us, and so we look for the moral of the tale. In other words, the shots which most clearly imply an authorial subject will tend to underline the social critique in the film, simply because that is the film's most serious line of narrative. This links up with the role of Dave, as the film's moral centre. It is his pronouncements which carry most weight and which are thereby linked to the authorial 'message.'

Camera movement in *Dead End* has the slightly contradictory character of combining some very obvious and notable movements with, overall, a fairly low amount of movement. Once again this indicates the ideological balancing act at work in the text. The ambiguity which camera movement can introduce is limited by the relatively low number of moving shots and by their conventional use to link scenes. In this way the conventional Hollywood balance is achieved, between movement to push the narrative on and stability to anchor the text. Yet the movements in *Dead End*, when they do occur, are often quite complex and thus introduce ambiguity as well as a slight feeling of instability, even if these meanings enter in a very limited way.

Thus the overall picture of the sub-codes of signification is of a complex pattern which allows the possibility of (and to some extent even encourages) a real conflict. Three ideological vectors can be seen working in the cinematographic style. One pushes the film in the direction of a social critique, following what is perhaps the strongest line in the narrative, another pushes the film back towards Hollywood norms thus defusing the more critical elements and a third introduces ambiguity and opens up the text, moving it away from any more specific direction.

SUBJECTIVITY

The point-of-view system in *Dead End* is fairly strong and coherent. The relatively large number of active characters makes identification with a group point-of-view easily achieved. Most shots are centred around either the kids or the main adult characters (Dave, Drina, Martin, Hunk, Kay and the doorman) with a further group appearing less often (the policeman, Martin's mother, Francey and Mr Griswald). Point-of-view is limited to these characters, either showing them looking or showing what they are looking at. Outside of these shots and encasing them are the establishing shots showing the whole scene and usually showing a character on whom the next shot will concentrate. This system creates a viewing subjectivity hierarchically linked to these characters. The overlapping of the different viewpoints brings the viewer into the text in a carefully ordered way.

The only breaks in this system are those shots in which, as already described, the enunciation is foregrounded. The extreme angled shots are the most obvious examples, but the opening and closing camera movements could also be mentioned. Although these shots tend to break the established subjectivity, they are 'neutralised' in various ways. Firstly (as noted in the earlier analyses) they serve as interpellative shots, inviting the view to adopt a particular position. This is particularly true of the opening shot. Secondly, they suggest another subject – not the viewing subject but the authorial subject. Unlike most of the breaks in subjectivity in *Dr Jekyll and Mr Hyde* which can be construed as a radical questioning of subjectivity, in *Dead End* the 'breaks' can all be contained within an authorial subject position which becomes conflated with a viewing position. Beyond the characters in the narrative, the viewer may be reminded of the 'storyteller' behind the text and thus an authorial position is suggested. This viewpoint is all-encompassing since the authorial position can function as a limiting position which includes the subjectivity on offer elsewhere in the text. To reject the offers of viewing position which the text makes is to reject its offer of pleasure. Alternative viewing positions produce either a lack of pleasure or a very different kind of pleasure from that which the Hollywood system tried to package. The authorial subjectivity becomes, then, the final guarantee of Hollywood's narrative pleasures.

CONCLUSION

The interest in what is happening here is twofold. Firstly, it shows how complex the ideological implications of style can be. This should not be too surprising since input to the film came from various sources – a left-wing play, the professionalism of Wyler, Toland and their collaborators, and the commercialism of Goldwyn, with none of these being strong enough totally to dominate the others. With the determinants of the film being, to a certain extent, contradictory, it is not surprising that the resultant text should also be contradictory.

The second point of interest is that it shows the direction taken by changes of style in the later thirties. The mixture of broadening sub-codes of signification and a subjectivity which is tighter than that found in many films in the early thirties is typical of the late thirties and early forties, especially after 1938. This will become more evident in the concluding chapters. At this point, *Dead End* appears as a film which, because of its production context, allowed the ambiguous expression of forces of change which would only be more generally apparent in 1939.

10 The End of the Decade

As the decade neared its end, major changes were underway, not just in the United States but throughout the world. This had immediate consequences, both in the American economy and in the representations of American life which Hollywood provided. Suddenly the enemy was not internal but external and very clearly seen. This resulted in changes in cinematographic style as well as content. *Dead End* provided an introduction to some of these changes, being an advance indicator of them due to its particular production context. But by 1939 the changes were becoming evident to any careful observer.

THE USA AT THE END OF THE THIRTIES

As before, the standard economic indicators will show the general trends. Statistics will be given up to 1942 to show the effect of the war.[1]

Year	*Unemployed*		*GNP*	*Economic growth*
	Number	*% Workforce*		
1938	10 390 000	19.1	84.7*	0.1**
1939	9 480 000	17.2	90.5	0.8
1940	8 120 000	14.6	99.7	1.3
1941	5 560 000	9.9	124.5	2.2
1942	2 660 000	4.7	157.9	2.9

* Gross national product is expressed in $billions by 1975 prices.
** Economic growth is calculated from a 1926 base.

The trends are clear from these figures. It is worth noting how fast unemployment was falling by 1940, although America did not enter the war until the end of 1941. Similarly GNP and economic growth show a dramatic improvement as the war started in Europe in 1939 (although of course this was before America's direct involvement in the fighting). Cinema statistics (which, as before, must be taken only as general indicators of a trend rather than precise data) do not show such a dramatic improvement but nevertheless reveal a healthy industry.

Year	Average weekly film attendance (millions)	Box-office receipts (millions of dollars)
1938	85	663
1939	85	659
1940	80	735
1941	85	809
1942	85	1 022

The main events of this period show a shift of emphasis from internal politics to international affairs.[2] The most important purely internal event was Roosevelt's re-election for a third term, defeating Wendell Willkie, although not by such a large majority as that by which he had defeated Alfred Landon in 1936. As far as foreign affairs are concerned, the crucial date is, of course, 7 December 1941 when Japanese aircraft destroyed most of America's Pacific fleet at Pearl Harbor in Hawaii. A simple view of American politics at this time would see an isolationist nation shocked into action by the Japanese attack. After all, the 1940 elections had seen both candidates promising not to involve the United States in any foreign wars. One of the most famous isolationists of the time, aviator Charles Lindbergh, was quoted in June 1940 as saying, 'Nobody wishes to attack us, and nobody is in a position to do so.'[3]

That such a view of American reactions is too simple, however, can be easily demonstrated. In January 1938 the Ludlow Amendment, which would have demanded a referendum before the country could go to war, was defeated, even if by a fairly small margin. In May of that year the Naval Expansion Act was passed, setting the country the aim of having a navy as strong as the joint forces of Germany, Italy and Japan. M. A. Jones notes how, when the war in Europe began in September 1939, Roosevelt, in contrast to Woodrow Wilson in 1914, 'pointedly refrained from urging the American people to be neutral in thought as well as in action'.[4] Preparations for war began in earnest in 1940. In June a National Defense Research Committee was created to organise research on new weapons (one of which was to be the atomic bomb) and in the same month Roosevelt announced that military aid would be given to Britain. In September the Selective Service and Training Act was passed, introducing conscription for the first time during peace. Then, in March 1941, the Lend–Lease Act was passed, allowing Roosevelt to give goods of all kinds to any country he saw as being important to America's defence.

The point to note about these events is that they show that, to all intents and purposes, American neutrality had been abandoned well before Pearl Harbor and Roosevelt was clearly in little doubt about America's inevitable involvement in the war. It is worth remembering that most of these developments required strong political support, at the very least in Congress in order to pass the Naval Expansion, Selective Service and Training, and Lend–Lease Acts. This is not to deny the isolationists' strength but by 1940 they were no longer the dominant voice they had been earlier in the decade.

In terms of ideology, the international situation both necessitated and enabled national cohesion. The approach of war took the United States nearer to unity than at any time since before the Wall Street Crash. At the same time, the improving economy reduced internal conflicts, allowing greater autonomy to such institutions as the film industry. If greater autonomy seems to conflict with the need for national unity, the answer can be seen to lie in the fact that the threat to unity came from outside the country. It was thus easier to unite the nation than it would have been if the threat had been internal, as it was seen to be by some at the beginning of the decade.

HOLLYWOOD REACTS

For Hollywood, then, the situation can be summed up as greater autonomy, as long as national unity of some kind was represented. In effect this created the mixture of a more adventurous style with a more nationalistic content. *Gone With the Wind* is the perfect example of this – a more varied style than that of the 1936–38 period is used in conjunction with an epic story of American history, showing the endurance and remaking of the nation after a period of crisis.

One result of the desire for national unity was the creation of an identifiable foreign enemy to unite the nation against. It was during the last years before the war that Hollywood finally acknowledged the trouble in Europe. Fear of losing money in other countries and of offending minority ethnic groups at home (such as the German-Americans) had kept explicit condemnations of fascism out of films during most of the thirties. (Earlier in the decade some films had even supported it – most famously *Gabriel Over the White House*, directed by Gregory La Cava at MGM in 1933). Those films that did refer to European politics did so in a very oblique fashion. *House of Rothschild* (Twentieth Century, 1934) condemned the ghettoisation of German Jews but was set in the early 19th century. As has already been noted, *Three Comrades* can be identified as anti-Nazi but only by the informed viewer. The one film of the period to refer to the Spanish Civil War was *Blockade*, a Walter Wanger production of 1938, directed by William Dieterle. However, the setting of the film is never identified and the politics of the unnamed civil war are reduced to a fight in defence of the local area against evil, militaristic invaders. In this context, the final speech of the film, in which the hero (Henry Fonda) addresses the camera directly, urging the audience's support, is a rather forlorn gesture. Only by a knowledgeable reading of the signs can *Blockade* be identified as a plea in favour of the Republican side in Spain.

By 1939, however, with war imminent in Europe, Hollywood began to take sides. Warner Brothers released their pseudo-documentary *Confessions of a Nazi*

Spy in April. In 1940 came *Foreign Correspondent* (a Walter Wanger production directed by Hitchcock), Chaplin's *The Great Dictator* and *The Mortal Storm* (MGM, d. Borzage). To a certain extent, all of these were exceptional. Hitchcock and Chaplin were English, and Borzage had already established his political sympathies, not just in *Three Comrades* but even as early as 1934 in *Little Man What Now*? Warner Brothers only made their anti-Nazi films three years after their German distribution arm had already been lost due to anti-semitism.[5] It was not until 1941 that the prospect of war began seriously to affect Hollywood film content (with Warner Brothers even distributing Harry Watt's *Target for Tonight* by the end of 1941).[6]

If the singling-out of a national enemy played only a minor role in films at this time, the same cannot be said for the recuperating emphasis on American history. 1939 was the year of the re-emergence of the Western as 'A' film material – *Stagecoach*, *Union Pacific*, *Destry Rides Again*, *Jesse James*, *Dodge City*, *Drums Along the Mohawk* are all historical films to a greater (*Union Pacific*) or lesser (*Destry Rides Again*) extent. Historical films in other genres include the musical *Hollywood Cavalcade*, the gangster film *The Roaring Twenties* and, of course, *Gone With the Wind*. There were also the biopics – *The Story of Alexander Graham Bell*, *Swanee River*, *Young Mr Lincoln* and *Abe Lincoln in Illinois*. Beyond these explicitly historical films are those which, in one way or another, glorify the American way of life, such as *The Wizard of Oz*, *Mr Smith Goes to Washington* and *Ninotchka*.

The greater autonomy mentioned earlier did have some effect on content, providing a contrast with the more nationalistically-minded films. 1939 saw the revival of the horror film with *Son of Frankenstein* appearing in January (the original film versions of *Dracula* and *Frankenstein* having been re-issued as a popular double bill in 1938).[7] 1939 was also the year which saw generic conventions being comically worked over in films such as *Destry Rides Again* and *The Cat and the Canary*. More interesting, perhaps, are those films which deal with situations which the earlier enforcement of the Production Code would have either banned outright or at least tried to put the audience's sympathies more unambiguously on the side of moral rectitude. Examples include *Intermezzo* (sympathy for the adulterer), *Bachelor Mother* (with the comedy centred around the hero's assumption that the heroine's baby is illegitimate) and *Daughters Courageous* (in which the father who abandoned his family for no real reason is treated sympathetically). Even the death from cancer of Bette Davis in *Dark Victory* is a far more downbeat ending (despite the heavenly chorus which accompanies her final moments) than would have been found in the middle of the decade. The heroine of *Three Comrades* dies from illness but, as already noted, the final scene is much more positive, and Garbo's death at the end of *Camille* is from a traditionally 'romantic' disease – consumption – and is safely set a hundred years earlier in a foreign country. The heroine of *Dark Victory* is the woman next door, dying from a very unromantic disease.

FILM STYLE AT THE END OF THE DECADE

The developments in cinematographic style can be summarised as a widening of the paradigms of signification but with the retention of fairly strong patterns of subjectivity. This mixture can be explained as the result of, on the one hand, the ideological need for unity, answered by strong subjectivity, moving towards a unified, national subject – the paradigmatic American viewer – and, on the other hand, the increased autonomy, allowing for the codes of signification to be broadened and the resulting expression of a certain degree of ideological tension and variation. By 1940 this 'certain degree' was still fairly limited. It is notable that the two main areas of stylistic change – lighting and camera angle – are arguably the two areas in which changes are least disruptive. The limited use of low-key lighting and unmotivated angles can easily function as a means of underlining the 'normality' of their conventional opposites. The darkness and obscurity present in some of these 1939 films is usually dispelled by the end, unlike the typical *film noir* in the mid forties which is liable to begin and end in darkness.

Yet, despite these limitations, the 'new' stylistic elements do have their effect. There is a complexity in this visual style which matches the ideological conflicts in many films of the period. Consider, for example, *Only Angels Have Wings*, directed by Howard Hawks at Columbia. At one level it is the typical Hawks celebration of macho values, in which women are seen as threats to the integrity of the male group. However this film takes these values almost to absurdity. The male characters are obsessive, foolhardy and reluctant to show any emotions. The lighting in the film is dark, with much use of shadow to separate and imprison characters. One effect of this is to emphasise the ambiguities of the male behaviour. The same film shot in high-key would have had a rather different effect.

At this point deep focus is relevant. Although it was a technique which was to play an important role in the development of the forties style, it is hardly noticeable in 1939, when Bert Glennon's cinematography in *Stagecoach* is one of the few obvious examples. Even the Wyler/Toland film of 1939 – *Wuthering Heights* – makes relatively little use of it compared with the same partnership's films of the early forties. However, it does fit into the pattern of stylistic change described above. Bazin's views on deep focus are well-known. According to him it 'reintroduced ambiguity into the structure of the image if not of necessity – Wyler's films are never ambiguous – at least as a possibility'.[8] In fact, what deep focus adds is not ambiguity but complexity. The increased autonomy of Hollywood in the late thirties and early forties allowed deep focus to be used and its specific uses serve to take films another step away from the simplicity and clarity of the more restrained paradigm of the middle of the decade.

The changes of 1939 become clearer when the lines of development are followed into the next decade. The deep-focus, long-take, mobile camera style of Orson Welles (*Citizen Kane*, 1941, *The Magnificent Ambersons*, 1942) and William Wyler (*The Letter*, 1940, *The Little Foxes*, 1941) and the related but distinct development of the *film noir* style with low-key lighting, frequent high and (especially) low angles, and often a fairly static camera (a style first appearing in *The Long Voyage Home*, d. Ford, 1940, *The Maltese Falcon*, d. Huston, 1941, and again *Citizen Kane*) are the well-known markers of the new developments. This account implies, incidentally, that the roots of *film noir* are firmly in the pre-war period. Paul Schrader has noted this, referring not just to a specific style, but also to content. 'Toward the end of the thirties a darker crime film began to appear (*You Only Live Once*, Fritz Lang, 1937; *The Roaring Twenties*, Raoul Walsh, 1939), and, were it not for the war, *film noir* would have been at full steam by the early forties.'[9] Rather than being a postwar phenomenon, as some critics have argued, *film noir* can be seen as a style which was developed before the war (remembering that for the USA it started in December 1941) and which the war years merely postponed. This suggests that rather than *film noir* being an aberration, it is rather the restrained style of the later middle thirties which now is aberrant, particularly when it is compared with the style of the early thirties (or even the range of styles available earlier in the closing years of the silent era or later in the fifties whose range can be seen by comparing the Arthur Freed MGM musicals with black and white 'realist' films such as *On the Waterfront* or *The Sweet Smell of Success*).

But for more precise evidence of the stylistic changes afoot in 1939 it is necessary to look in more detail at the films themselves. In order to illustrate the changes beginning to take place at this time, two films will be examined – a generic detective film, *The Hound of the Baskervilles*, and the most famous film of the year – *Gone With the Wind*.

THE HOUND OF THE BASKERVILLES

Made at Twentieth Century-Fox and premiered in March 1939, *The Hound of the Baskervilles* was directed by Sidney Lanfield, with cinematography by Peverell Marley. The pairing in this film of Basil Rathbone as Sherlock Holmes and Nigel Bruce as Dr Watson was popular enough to lead to *The Adventures of Sherlock Holmes* later in that same year and then to a series of twelve more films made at Universal between 1942 and 1946 until Rathbone refused to do any more.[10] The spawning of a series of sequels is perhaps a clearer indication of the popularity of the original than any box-office figures could be.

Cinematographic style in *The Hound of the Baskervilles* is not greatly different from the restrained style except in the use of darker lighting schemes and unmotivated camera angles. The film opens with a scene showing the death of Sir Charles Baskerville. The setting is a typical 'blasted heath', familiar from many horror films. Behind the opening titles is a tracking shot over a dark moor, ending with a view of Baskerville Hall at night. The next seven shots are as follows. During the first four, the sound of a hound is heard.

Shot 1 A man is running towards the left of the frame, moving from ELS to LS. There is some fairly light shade in the background and darker shadows of rocks in the middle ground and foreground. The man is in light shadow.

Shot 2 The man is in MS as the camera tracks with him. There is some shadow on his face. The background is in light shade.

Shot 3 A pan follows the man from LS to MLS. Again there is light shade in the background and some shadow on his face with foreground silhouettes of rocks and trees.

Shot 4 The man reaches gates. In the background beyond the gates there is a pool of diffuse light. The middle ground is darker and there is a strong silhouette of a tree in the left foreground with undefined shadow in the right foreground. The man runs through the gates into MLS, the camera panning and tracking with him. He collapses.

Shot 5 Another man (later revealed to be an escaped convict, the Notting Hill murderer) appears in MCU with scattered shadow on his face. The right quarter of the frame is completely dark. The strongest light is in the background.

Shot 6 The collapsed man lies in the foreground. The convict comes up to him and kneels at his body but runs off when a voice is heard. There is some light in the background and a little at the head of the collapsed man, but most of the frame is in light shade.

Shot 7 The shot begins with a completely dark frame which is then revealed to be a door when it opens. A woman (Mrs Barryman, the housekeeper) appears in MLS. A little light can be seen in the room behind her, but the foreground is dark and her face is partially concealed. She walks into MS holding a lamp. Her face becomes fully visible. She sees something and screams. Fade out.

The characteristic lighting set-up in this scene is to have a pool of light in the background, and the rest of the set in medium shade, with some stronger shadows and silhouettes around the sides of the frame and in the background. Although the contrasts are not high and the light is often diffuse, the dominant tones are fairly dark.

In the scenes in which Dr Watson and Sir Henry Baskerville (Richard Greene) first find Barryman (John Carradine) signalling to the convict on the moor and then go out on the moor themselves (all of which takes place at night), there are much stronger shadows. Some are on walls (as Watson and Sir Henry come down the stairs) and some are on faces, sometimes completely concealing the features (Watson at the window after he blows out the candle, both Watson and Sir Henry when they are first seen on the moor). Some shadows are caused by strong, unusually placed lights (such as the shots of the convict as he watches Watson and Sir Henry, in which the lights are low and to the side). Similarly, in the seance scene, strong low-placed key lights create a series of powerfully shadowed images, most of them in CU.

These lighting effects are unremarkable in themselves – they are, after all, used in scenes of high tension in a genre which frequently uses such devices. However, their relevance here is to emphasise their near-absence from mainstream studio films in the immediately preceding years.

Slightly more unusual is the use of unmotivated angles. There are not many of these in the film but there are enough to be worthy of comment. Three examples will indicate their use. The first is in Holmes' rooms. Watson, Sir Henry and Dr Mortimer (Lionel Atwill) are discussing with Holmes the threatening note which Sir Henry has just received. Apart from the closing shot of the scene, only one shot shows all four characters – Watson sitting in the left foreground, Mortimer standing in the background slightly to the right of Watson, Holmes sitting in the middle of the background and Sir Henry sitting on a table in the right foreground. The camera is at knee level, looking up at the four characters. A more extreme angled shot occurs much later in the film, at the beginning of the scene in which Holmes reveals to Barryman and his wife that the convict is dead. The shot begins with the camera at thigh height looking up at Barryman, Watson and Holmes in MLS. Watson walks out of the frame and the camera moves forward to get Barryman and Holmes in MS, at an even steeper angle. Finally, as an example of an unmotivated high angle, there is the fairly brief shot of Holmes and Watson from above their heads as they rescue Sir Henry from the hound.

These examples of lighting and camera angle are by no means extreme. However they do indicate how, even in a fairly cheaply-made generic film, the stylistic paradigm was broadening by 1939. Scenic structuring and the construction of subjectivity, on the other hand, remain well within the norms of the 1936–38 period.

The detective story has a long and well-known history and the narrative structure in which a central male character uses superior intellect and skill to solve a complex crime and capture a life-threatening criminal has been common in Hollywood since the earliest days of film. Even the Sherlock Holmes stories themselves have been common on screen for many years with American, Danish, French, German and British versions all appearing before the outbreak of the First World War. The broad ideological pattern of such films will be clear – strongly sexist, emphasising the power of human individuality, accepting the natural authority of those who know better and based in a sense of natural morality – as will the strength of the narrative which pushes towards the revelation of a mystery and concludes with complete closure and no loose ends. But such conventions should not obscure the differences between the various fictional detectives or their various filmic manifestations. If the film of *The Hound of the Baskervilles* is compared with the printed story then the former will be found to be even more regressively sexist than its original. In Conan Doyle's story, Stapleton's sister is a more impressive character than she appears in the film (and is finally revealed to be not his sister but his wife and to some extent a partner in crime) and the stronger character of Mrs Laura Lyons is completely omitted from the film. However when compared with other screen detectives of the middle and later thirties, such as Charlie Chan and Mr Moto (both film series being made at Twentieth Century-Fox, like the first Holmes films) Sherlock Holmes is less conventional. Both Chan (played by Warner Oland during most of the thirties, until Sidney Toler took over in 1938) and Moto (played by Peter Lorre, 1937–39) are uncomplicated characters, essentially given colour by their oriental origins. Holmes on the other hand is a much odder character, with his obsessive moods, violin-playing and drug-taking (this last being such a well-known feature of the character that it is even briefly referred to in the 1939 film, the last line of which, uttered by Holmes as he retires to bed, is 'Oh, Watson, the needle!'). The broadened norms of lighting and camera angle in *The Hound of the Baskervilles* lead to a film much darker and more expressive in mood than the earlier Chan and Moto films, thus allowing the less respectable aspects of Holmes to be emphasised and given depth. He becomes a dark man in a dark world, despite the evident respectability and simplicity of such characters as Dr Watson and Sir Henry Baskerville who merely serve to point up the contrast with the detective. This does not constitute a major change or disturbance of ideology, but it does take the film a step away from the certainties and simplicities of the previous years as the ideological structuring of the film is slightly deflected by the cinematographic style – a feature which is evident in a number of other films in 1939.

GONE WITH THE WIND

Although *The Hound of the Baskervilles* was a popular film, it did not get into the top ten box-office hits. That stylistic change was occurring in the most popular and expensively-made films can be shown by looking at the film which, for many people, sums up the classic Hollywood period – *Gone With the Wind.* The background to this film is well-known and has been described in full by Roland Flamini.[11] Although premiered right at the end of the decade (December 1939), pre-production began in 1936 with shooting beginning in December 1938. The large number of people involved in making the film is important since it emphasises that it was not an aberration, created in the main by one man's activity. Its making involved many of the major talents in thirties Hollywood. Producer David O. Selznick was not just a businessman – he was frequently on the set, suggesting camera set-ups and even directing a few shots. He also supervised the editing. Flamini has estimated that the various directors contributed as follows: Victor Fleming (the officially credited director) 55 per cent of the final film, Sam Wood (who took over after Fleming collapsed with exhaustion) 15 per cent, George Cukor (the original choice of director who, after being removed from the production, continued to act as an unofficial adviser to Vivien Leigh and Olivia de

Havilland) 5 per cent, 1 per cent by second-unit specialist B. Reeves Eason, 15 per cent by William Cameron Menzies (on the credits he is given an unusually prominent mention stating 'this production designed by William Cameron Menzies') and the remaining 9 per cent being optical effects and titles. The cinematography is credited to Ernest Haller with Ray Rennahan and Wilfrid Cline from the Technicolor Corporation, although Flamini notes that Lee Garmes was the original cinematographer and shot what was to become the first hour of the three-and-three-quarter hour film.

The success of *Gone With the Wind* is legendary.[12] It won Academy Awards for Best Picture, Best Director, Best Actress (Vivien Leigh), Best Supporting Actress (Hattie McDaniel), script (Sidney Howard), colour cinematography, interior decoration (Lyle Wheeler) and editing (Hal Kern), as well as getting nominations for Best Actor (Clark Gable) and another Best Supporting Actress (Olivia de Havilland). In addition to all these there was an award for 'pioneering in the use of coordinated equipment' and special awards for Technicolor ('for bringing three-color feature production to the screen') and Menzies ('for outstanding achievement in the use of color for the enhancement of dramatic mood in the production *Gone With the Wind*'). To top it all, Selznick received the Irving Thalberg Memorial Award.[13] All that remained to be achieved after this was box-office success and that was not long in coming. Flamini notes that by the end of 1940 it had been seen by 25 million people in the United States alone and had earned over $14 million, with the final cost of the production having been just under $4 million.[14] It was a success economically, technically and aesthetically.

The first aspect of the cinematographic style to be noted is the use of more extreme lighting effects than the 1936–38 paradigm allowed for, in particular the use of silhouettes. These occur at several points of major importance. Behind the opening titles (which last for almost four minutes, from the studio logo to the end of the introductory scene-setting titles) there are various shots of an idyllic Old South. Most are conventional landscapes or close shots of magnolia blossom. However, the sequence ends with a low-angle shot of a sky-line silhouette of slaves herding cattle. The only colour in the shot is in the sky itself. Not far into the film there is a scene in which Scarlett O'Hara (Vivien Leigh) is speaking with her father (Thomas Mtichell) and being told by him of the importance of land in general and Tara in particular. The scene ends with a low-angle silhouette LS of the two characters before the camera pulls back to ELS with a silhouetted tree to the left of the frame and Tara in the distance.

Another prominently-placed example is in the scene in Atlanta when Melanie (Olivia de Havilland) is about to give birth, helped only by Scarlett and Prissy the maid (Butterfly McQueen). The relevant sequence is the following.

Shot 1 Scarlett in MS, with Melanie's head visible in the bottom right corner of the frame and Prissy moving in the background. There are shutters on the window. Everything is in silhouette.

Shot 2 CU of Melanie lying back in silhouetted profile.
Shot 3 ECU of Scarlett in silhouetted profile.
Shot 4 Repeat of the set-up in shot 1.

Later, when Scarlett and Prissy arrive at Twelve Oaks after the war, their passing of the house sign is preceded by an ELS of the two women in horse and cart, with a burnt-out building on the left of the frame and a tree on the right. All is in silhouette against an orange sky. The first half of the film ends with Scarlett walking out onto the land at Tara and swearing, 'I will never be hungry again.' There are four shots in this scene.

Shot 1 ELS of Scarlett walking out past a fence, all in silhouette against an orange sky.
Shot 2 CU of her hand grasping in the soil, in strong shadow but not a silhouette.
Shot 3 MLS of Scarlett. She falls to the ground and then gets up, and the camera moves into CU. This shot is very dark, in virtual silhouette at the beginning but with just enough light at the end to make out her facial features.
Shot 4 LS of Scarlett. The camera moves back to ELS. A tree is at the side and a fence crosses the frame. All is in silhouette.

This last shot is clearly related to the silhouette of Scarlett and her father mentioned earlier but an even closer variation occurs at the very end of the film with the same set-up (tree to left, Tara in the distance) but with only Scarlett in the centre. Over this the words 'The End' appear.

These silhouettes are among the most striking visual effects but there are also notable scenes in low-key lighting. Just after Ashley (Leslie Howard) returns to the war there is a title beginning, 'Atlanta prayed while onward surged the triumphant Yankees.' Behind this title and continuing after it is a sequence-shot (lasting 77 seconds) of Melanie and Scarlett in a hospital. It begins with a view of a stained-glass window and then the camera pans round to reveal two shadows on a wall. At this point the titles disappear. The pan continues and reaches Melanie and Scarlett in MS in the foreground with their shadows still dominating the background. The camera then moves into MCU and rises above the women's heads to end with another view of the two shadows.

The longest, and most famous, dark sequence is the escape from Atlanta. Rhett (Clark Gable), Scarlett, Melanie with her baby, and Prissy ride through Atlanta in a horse and cart while the town burns, before joining the flow of refugees. This sequence (which, timed from the point at which the main characters leave their town house up to Rhett's departure, lasts for almost eight-and-a-half minutes) includes frequent facial shadow and foreground shadow. It is climaxed by extreme low-key lighting on Rhett and Scarlett as he leaves her to join the Confederate army. Some of the shots at this point are in virtual silhouette, with only a rim of light round the faces, all against an intense red sky.

A little later, after Scarlett has reached Tara, there is another extreme low-key scene in which, while talking to her father, she realises that his mind has gone. Again a virtual silhouette with a rim of light is used, with other shots in the scene having faces either half-lit or lit completely in low light but surrounded in dark shadow. A brief scene with the servants, also in low-key, leads into the silhouette scene (already described) which ends the first part of the film.

The second part of the film has less low-key but the scene in which Rhett comforts his daughter in London is very darkly lit and the final confrontation between Rhett and Scarlett, after Melanie's death is, if not in low-key, certainly on the dark side of mid-key, with contrasting light shadows frequently seen on faces.

As striking as the lighting is the use of unmotivated angles. In some of the scenes already mentioned these occur, particularly low angles, for example the silhouette of Scarlett and her father at Tara, and some shots in the sequence of the burning of Atlanta. There are a number of other unmotivated low angles but the most extreme, and the most striking, occur during the bombardment of Atlanta as the panic starts. Scarlett sees some of her family's servants marching out to dig trenches and goes to speak to them. When she bids them goodbye the camera is virtually at ground-level. As she leaves the frame, horses gallop across the foreground. After a brief MCU of Scarlett, the low-angle set-up is repeated but without her in view.

If this is the most extreme low angle, the most extreme high angle must be the well-known crane shot of the dead and wounded in the railway station in Atlanta. Scarlett emerges into the scene, looking for Dr Meade. There is a CU of her looking around, and then the crane shot begins with Scarlett in LS. The camera moves slowly back and up, revealing the vast number of men lying around, until Scarlett is almost lost among them. In the foreground a tattered Confederate flag appears. The shot lasts for 54 seconds. It is remarkable partly for its height (Flamini notes that the camera ended up 90ft above the ground)[15] but also for the way it loses its central character. Scarlett becomes a difficult figure to find by the end of the shot, which radically departs from anything approaching her point-of-view – at the end she is still looking for the doctor. Although clearly a version of the establishing shot which opens on a detail and then moves back to reveal the whole scene, this is more extreme than the typical mid-thirties examples.

This leads on to camera movement, which does not depart so far from the restrained style as the use of lighting and camera angle. There are a number of travelling shots which are unusual or at least involve a bit more movement than might have been expected. One example concerns Rhett's first appearance. Scarlett and a friend stop on the stairs at Twelve Oaks and Scarlett asks who the stranger is. The next shot looks down the stairs at Rhett. The previous set-up is then repeated as her friend tells Scarlett about him. The fourth shot of the sequence starts with the view down the stairs but then moves towards him, from LS to MS, although none of the three characters has moved. (A similar movement occurs in another well-known film of 1939, *Stagecoach*, when the Ringo Kid (John Wayne) first appears and the camera moves without character motivation from MLS to

CU). Other movements in the film are like the high-angle crane shot – conventional uses of camera movement but slightly emphasised. A notable example in the second half of the film is the shot which introduces the chain-gang labourers who will be used by Scarlett in her timber yard. The camera begins at CU on a timber saw and then moves down to show the chained legs of the prisoners and then up to show their faces, finally panning round to show Scarlett in the background. This 27-second shot also includes wall shadows, foreground shadows and near-silhouettes.

It will already be clear that *Gone With the Wind* makes much use of very distant ELSs with human figures dwarfed in a landscape. Very close shots are not so common, although ECUs are found in the scene in which Scarlett is almost raped by a carpetbagger, and in the following scene there is an eyes-only ECU of Scarlett as she and the other women await the return of the men who have set out to avenge her.

If this seems to be a long list of stylistic variation in one film (and it is certainly not a complete list), the question will arise as to how these variations interact with the content. Three points can be made initially. The first is that *Gone With the Wind* lasts for three hours and forty-five minutes and so the scenes described above are only a small part of the overall film. They thus represent only a slight widening of the style, rather than a major transformation such as would appear in the classic *noir* films of the forties. The second point is that the content of the film seems to be ideologically 'safe', certainly at an obvious level – it is historical, heterosexual romance, strongly patriarchal and racist, and concerned with one of the formative myths of modern America (showing the South as not all bad and the North as not all good, thus evening up the score and unifying the nation). The third point to make is that many (although not all) of the scenes mentioned are conventional settings for an 'expressive' treatment – the dark night of the Atlanta siege, and scenes of birth, madness and death.

Despite this, it is worth pointing out that *Gone With the Wind* is not as ideologically cohesive as its role in film history might suggest (in fact its enduring popularity may be partly because of this). It does not accomplish perfectly the difficult task of romanticising the Old South. Slavery cannot be totally repressed from any film dealing with the Civil War and the silhouette which closes the opening titles can strike an ominous tone which, for some viewers, may never be completely banished. More importantly, however, the film allows for a feminist reading or, at least, it allows for a reading which conflicts with the dominant patriarchal tone. Although Scarlett is the leading character – she is seen far more than Rhett and the story is essentially *her* story – the 'wisdom' of the film is carried by Rhett. It is his comments (on war, on Scarlett, on the Old South) which are given the weight of the film's 'truth'. His judgements and his actions are never brought into doubt, the way that those of all the other main characters are (Melanie may be presented as a supremely good character but she does not see all that is going on around her in the way that Rhett does – her comments may be proved wrong by ignorance; Rhett's never are). Despite this weighting of Rhett, the film remains the story of one very strong-willed independent woman who scorns conventions, survives the war virtually unaided and rebuilds her life and career by herself. From one point of view, the film ends with Scarlett abandoned by the one man she has ever really loved. From another point of view, it ends with Scarlett once more independent and looking forward – even if with a somewhat forced optimism – to 'another day' and the final shot sees her in command of the family estate, taking the position which her father had at the beginning of the story. There is, then, a certain ideological instability in the film, which the stylistic variations can be read as emphasising. It is no accident that most of the scenes described above include Scarlett.

It would be rash, however, to read too much into the style of specific scenes. More important is the overall stylistic system and the crucial point is that the system of *Gone With the Wind* is, in significant respects, different from the prevailing style of the preceding period. It is the connotation of the style as a whole which sets this film – and many others of the 1939–41 period – apart from the earlier films. The style is less limited, less stable, less obvious in what it shows, less predictable and less ideologically secure.

CONCLUSION

The end of the decade, then, sees cinematographic style beginning to develop along lines which would become more obvious during the forties. Unlike the variations apparent in the early thirties, however, strong subjectivity remains. As noted earlier, subjectivity is the most important aspect of cinematographic style as far as ideology is concerned and this should make clear the difference between the films of the early thirties and those of the late thirties. In the earlier period ideological crisis was having its effect, in the later period ideological unity was to the fore. In both periods Hollywood's position of relative autonomy was making its products into complex effects of ideological conditions, reflecting in some way not just the dominant ideology but also its gaps, tensions and confusions as well as, occasionally, other ideologies and variations of the dominant ideology. To read 'the classic Hollywood film' of the thirties (and of course the preceding chapters should make it clear that this term is itself a mythic and therefore ideological construction) as an ideologically monolithic and stable representational form is to force a particular reading on to texts which do not in themselves support such a reading.

Conclusion

The preceding chapters have described a series of changes in Hollywood cinematographic style during the thirties. The decade begins with a fairly open, adventurous style with wide contrasts of lighting, angle and distance, direct-to-camera looks and frequently unpredictable structuring. There is then a transitional period from 1933 to 1935 in which the style becomes more restrained, reaching its narrowest paradigm between 1936 and 1938. This style is dominated by high-key lighting, motivated angles and movement, predictable structuring and unproblematic subjectivity. In 1939 the style begins to open up again but mainly in the areas of lighting and angle and, to a lesser extent, movement.

All these statements are, of course, generalisations. Just as there are films made in 1932 which are in a restrained style, so there are occasional films in the 1936–38 period which go beyond the contemporary norms (*Dead End* being an obvious example). Most of these are films made in exceptional circumstances – for example independent productions by Goldwyn, Selznick and Wanger, or films made by directors who were in a position to, in effect, create their own conditions. The most obvious example is John Ford, not just in the films he made as a guest director at RKO but also in some of the films he made as a contract director at Twentieth Century-Fox. *Four Men and a Prayer*, made in 1938, includes plentiful use of very dark shadow, often covering characters' eyes. Just occasionally a more routinely-produced film shows signs of such 'deviation'. *Seventh Heaven* was directed by Henry King at Twentieth Century-Fox in 1937 and includes canted angles used not to add tension to an action scene (as happens, for example, in *The Adventures of Robin Hood* and *The Charge of the Light Brigade*) but in a more conventional scene. This should emphasise the point that these descriptions of style are merely generalisations, and given the varieties of ideological pressures working on any single studio-made film, such apparent 'lapses' should be unsurprising.

The relation of the codes and sub-codes of cinematic style to ideological determinations is, of course, not a hard and fast scientific rule but a deduction of likely links from circumstantial evidence using a coherent theoretical framework. The changes in the American economic, political and social fabric, as the nation moved from the Wall Street Crash and the depths of the Depression through the New Deal years and finally towards international conflict, provide a background which a theory of ideology can link to the stylistic changes seen in the contemporary films. It should be clear that the argument of this study has not been simply to make direct links between changes in the economy and changes in film style. The latter changes were influenced by a number of factors and do not simply reflect society in any one-to-one way (indeed it is not clear how cinematographic style *could* reflect society in any simple way). This is most obvious with the emergence of the full restrained style in 1936 but is also seen in the variations evident in 1939.

This, then, concludes the argument. It is worth repeating at this point that the concern has been with the causes of cinematographic style, not the effects. There is a natural temptation to extend the argument to reach conclusions about the audience and so to make assertions about the ideological effects of Hollywood films in the thirties. As noted earlier, the empirical evidence needed to support such conclusions is not available. It was also pointed out that films cannot force the viewer into a particular position – they can only suggest that position, promising pleasure if it is adopted. To add to the complications, it was also noted that the audience was not a homogeneous whole, making it even more likely that major differences in response would have occurred. Media effects are, of course, a notoriously difficult area in which to achieve any degree of certainty and it is arguable that the principal effects are of reinforcement and agenda-setting, rather than persuasion, making them even more difficult to identify. A final point is that film viewing is not likely to be the only media consumption by the audience. In the United States during the thirties, radio was very popular, with shows such as the *Lux Radio Theatre* getting audiences of around 30 million.[1] To consider the effects of film-viewing alone is misleadingly to isolate one element of the cultural experience of the audience. This will bring a false simplicity to a very complex situation.

Speculations about possible audience effects must, then, be rejected. What remains of the argument can be compressed into three points. The first is that cinematographic style can be related to ideological determinants. The second is that there were significant stylistic changes in Hollywood films during the thirties. The third is that ideology is a necessary element in any adequate explanation of these changes. The previous ten chapters have attempted to demonstrate these points.

Notes

1 INTRODUCTION

1. C. Higham, *Hollywood Cameramen* (London: Thames and Hudson, 1970), p. 10.
2. Ibid., p. 11.
3. A. Sarris, *The American Cinema* (New York: E. P. Dutton, 1968), p. 31.
4. L. Maltin, *The Art of the Cinematographer* (New York: Dover, 1978), p. 12.
5. Ibid., p. 17.
6. Ibid., p. 19.
7. Ibid., p. 26.
8. T. Rainsberger, *James Wong Howe, Cinematographer* (London: Tantivy Press, 1981), p. 61.
9. Ibid., p. 162.
10. R. Williams, *Television: Technology and Cultural Form* (Glasgow: Fontana, 1974), p. 130.
11. P. Ogle, 'Technological and Aesthetic Influences on the Development of Deep Focus Cinematography in the United States', in B. Nichols (ed.), *Movies and Methods*, Vol. 2 (London: University of California Press, 1985), p. 64.
12. A. Bazin, *What is Cinema?*, Vol. 1 (London: University of California Press, 1967), pp. 21 and 30.
13. Ibid., p. 21.
14. D. Bordwell, J. Staiger and K. Thompson, *The Classical Hollywood Cinema* (London: Routledge and Kegan Paul, 1985), p. 244.
15. Ibid., p. 7.
16. Ibid., p. 6.
17. N. Burch, *Life to Those Shadows* (London: British Film Institute, 1990), pp. 260–1.
18. N. Burch, *In and Out of Synch* (Aldershot: Scolar Press, 1991), p. 165.

2 IDEOLOGY AND CINEMATOGRAPHIC STYLE

1. The following section is based on Chapter 1 of the author's *Ideology* (London: Batsford, 1992) where the definition is developed more fully. The three texts mentioned earlier are: J.-L. Comolli and J. Narboni, 'Cinema/Ideology/Criticism,' in B. Nichols (ed.), *Movies and Methods* (London: University of California Press, 1976), pp. 22-30; P. Biskind, *Seeing Is Believing* (London: Pluto Press, 1984); B. Nichols, *Ideology and the Image* (Bloomington: Indiana University Press, 1981).
2. J. Collins, *Uncommon Cultures* (London: Routledge, 1989), p. xiv.
3. The original arguments can be found in N. Abercrombie, S. Hill and B. S. Turner, *The Dominant Ideology Thesis* (London: Allen and Unwin, 1980) with the updated version in N. Abercrombie, S. Hill and B. S. Turner (eds), *Dominant Ideologies* (London: Unwin Hyman, 1990).
4. T. Eagleton, *Ideology: An Introduction* (London: Verso, 1991), p. 45.
5. J. B. Thompson, *Ideology and Modern Culture* (London: Polity Press, 1990), p. 56.
6. Ibid., p. 57.

7. See Chapter 8 of the author's *Ideology* for an example of a general ideological analysis of a recent film (*Raging Bull*, d. Martin Scorsese, 1980).
8. R. Barthes, *S/Z* (New York: Hill and Wang, 1974), p. 20.
9. J. B. Thompson, op. cit., p. 275.
10. K. Silverman, *The Subject of Semiotics* (Oxford: Oxford University Press, 1983), p. 270.
11. R. Barthes, op. cit., p. 100.
12. F. Basten, *Glorious Technicolor* (London: Thomas Yoseloff, 1980), pp. 170–1.
13. C. Metz, *Language and Cinema* (The Hague: Mouton, 1974), pp. 28–9.
14. Ibid., p. 62.
15. Ibid., p. 63.
16. Ibid., pp. 75–6.
17. Ibid., p. 86.
18. S. Russell, *Semiotics and Lighting* (Ann Arbor: UMI Research Press, 1981), pp. 44–5.
19. B. Salt, *Film Style and Technology* (London: Starword, 1983), pp. 389–91.
20. Ibid., p. 243.
21. J. Paine, *The Simplification of American Life* (New York: Arno Press, 1977), pp. 210–11.
22. Ibid., p. 16.
23. S. Neale, *Genre* (London: British Film Institute, 1980), p. 55.
24. S. Heath, *Questions of Cinema* (London: Macmillan, 1981), pp. 19–75.
25. D. Bordwell, *Narration in the Fiction Film* (London: Methuen, 1985), p. 25.
26. J. Feuer, *The Hollywood Musical* (London: Macmillan, 1982), p. 36.
27. E. Branigan, *Point of View in the Cinema* (Berlin: Mouton, 1984), p. 103.

3 FROM SILENT TO SOUND: *ALL QUIET ON THE WESTERN FRONT*

1. D. Gomery, *The Hollywood Studio System* (London: Macmillan, 1986), p. 150.
2. C. R. Barker and R. W. Last, *Erich Maria Remarque* (London: Oswald Wolff, 1979), pp. 15 and 165.
3. Ibid., pp. 19–20.
4. Details of the production history in the following paragraph are taken from C. Denton, K. Canham and T. Thomas, *The Hollywood Professionals, Vol. 2: Henry King, Lewis Milestone, Sam Wood* (London: Tantivy Press, 1974).
5. A. Estrin, *The Hollywood Professionals, Vol. 6: Capra, Cukor, Brown* (London: Tantivy Press, 1980), p. 97.
6. Slide's claim can be found in his article '*All Quiet on the Western Front*,' in C. Lyon (ed.), *The Macmillan Dictionary of Films and Filmmakers*, Vol. 1 (London: Macmillan, 1984), p. 22. The information on *Broadway* comes from Barry Salt's *Film Style and Technology* (London: Starword, 1983), p. 228.
7. S. Eisenstein, *Film Form* (London: Harcourt Brace Jovanovich, 1949), p. 75.
8. C. Denton *et al.*, op. cit., p. 76.
9. K. Thiede, 'Lewis Milestone: a Film Checklist', in J. Tuska (ed.), *Close Up: the Contract Director* (Metuchen: Scarecrow Press, 1976), p. 347.
10. H. J. Forman, *Our Movie Made Children* (New York: Macmillan, 1933), p. 254.
11. Details of Academy Awards and *New York Times* rankings are taken from C. Steinberg's *Reel Facts* (Harmondsworth: Penguin, 1981).
12. C. Denton *et al.*, op. cit., pp. 79–80.

13. D. Bordwell, J. Staiger and K. Thompson, *The Classical Hollywood Cinema* (London: Routledge and Kegan Paul, 1985), pp. 228–9.
14. Ibid., p. 229.
15. Ibid.
16. B. Salt, op. cit., pp. 222–3.
17. Ibid., p. 265.
18. Ibid.

4 A CRISIS OF EXPLANATION: THE EARLY THIRTIES

1. Statistics are taken from the US Bureau of the Census, *Historical Statistics of the United States* (Washington: Bureau of the Census, 1975), pp. 126 (unemployment), 224 (GNP), 226 (economic growth) and 400 (film statistics).
2. W. E. Leuchtenburg, *The Perils of Prosperity, 1914–32* (London: University of Chicago Press, 1958), p. 261.
3. M. A. Jones, *The Limits of Liberty* (Oxford: Oxford University Press, 1983), p. 457.
4. Ibid., p. 467.
5. Ibid.
6. J. Major, *The New Deal* (London: Longmans, 1968), pp. 47–8.
7. M. A. Jones, op. cit., p. 456.
8. Ibid., pp. 455–6.
9. Ibid., p. 456.
10. D. Bordwell, J. Staiger and K. Thompson, *The Classical Hollywood Cinema* (London: Routledge and Kegan Paul, 1985), pp. 314–15.
11. B. Salt, *Film Style and Technology* (London: Starword, 1983), p. 265.
12. D. Bordwell *et al.*, op. cit., p. 343.

5 QUESTIONING SUBJECTIVITY: *DR JEKYLL AND MR HYDE*

1. Unless otherwise noted, all details of Paramount's history are taken from Chapter 2 of Douglas Gomery's *The Hollywood Studio System* (London: Macmillan, 1986).
2. D. Bordwell, J. Staiger and K. Thompson, *The Classical Hollywood Cinema* (Routledge and Kegan Paul, 1985), pp. 320–2.
3. Ibid., p. 328.
4. H. T. Lewis, *The Motion Picture Industry* (New York: Jerome S. Ozer, 1971), p. 30.
5. D. Bordwell *et al.*, op. cit., p. 328.
6. H. T. Lewis, op. cit., p. 352.
7. B. Schulberg, *Moving Pictures* (Harmondsworth: Penguin, 1984), p. 359.
8. Mamoulian's history taken from Tom Milne's *Rouben Mamoulian* (London: Thames and Hudson, 1969).
9. J. Calder, *RLS: a Life Story* (London: Hamish Hamilton, 1980), p. 230.
10. S. S. Prawer, *Caligari's Children* (Oxford: Oxford University Press, 1980), pp. 86–7.
11. C. Steinberg, *Reel Facts* (Harmondsworth: Penguin, 1981).
12. H. J. Forman, *Our Movie Made Children* (New York: Macmillan, 1933), pp. 99 and 110.
13. H. T. Lewis, op. cit., p. 388.
14. S. S. Prawer, op. cit., p. 86.

15. Quotations from *The Strange Case of Dr Jekyll and Mr Hyde* are taken from R. L. Stevenson, *The Collected Shorter Fiction*, ed. by Peter Stoneley (London: Robinson, 1991), pp. 439–88.
16. D. Robinson, 'Painting the Leaves Black', *Sight and Sound*, 30, 3 (Summer, 1961), p. 125.
17. B. Salt, *Film Style and Technology* (London: Starword, 1983), pp. 261 and 282.
18. Ibid., p. 245.
19. T. Milne, op. cit., p. 45; P. Lehman, 'Looking at Ivy Looking at Us Looking at Her', *Wide Angle*, 5, 3 (1983), pp. 59–63; S. S. Prawer, op. cit., p. 91; D. Bordwell, *Narration in the Fiction Film* (London: Methuen, 1985) p. 350, footnote 26.
20. D. Bordwell, ibid.
21. D. Bordwell, *et al.*, op. cit., p. 305.
22. Ibid., p. 304.
23. Ibid., p. 228.
24. Ibid., p. 10.
25. Ibid., p. 229.
26. R. Dyer, *Stars* (London: British Film Institute, 1979), p. 30.

6 THE NEW DEAL IN HOLLYWOOD, 1933–35

1. Statistics are taken from the US Bureau of the Census, *Historical Statistics of the United States* (Washington: Bureau of the Census, 1975) pp. 126 (unemployment), 224 (GNP), 226 (economic growth) and 400 (film statistics).
2. M. A. Jones, *The Limits of Liberty* (Oxford: Oxford University Press, 1983), p. 461.
3. D. K. Adams, *Franklin D. Roosevelt and the New Deal* (London: The Historical Association, 1979), p. 25. Details concerning the New Deal administration in the following paragraphs are taken from M. A. Jones's *The Limits of Liberty* and William Miller's *A New History of the United States* (London: Paladin, 1970).
4. M. A. Jones, op. cit., p. 464.
5. Ibid., p. 474.
6. Ibid., p. 475.
7. D. K. Adams, op. cit., p. 8.
8. S. Terkel, *Hard Times* (London: Allen Lane, 1970), p. 135.
9. Ibid., p. 193.
10. M. A. Jones, op. cit., p. 462.
11. Details of the Payne Fund studies are taken from Garth Jowett's *Film: the Democratic Art* (Boston: Little, Brown and Company, 1976), pp. 220 and 230–1.
12. H. J. Forman, *Our Movie Made Children* (New York: Macmillan, 1933), p. vii.
13. G. Jowett, op. cit., p. 226.
14. H. J. Forman, op. cit., p. 272.
15. R. Moley, *Are We Movie Made*? (New York: Macy-Masius, 1938), pp. vii and viii.
16. Ibid., p. 61.
17. Ibid., p. 63.
18. M. Quigley, *Decency in Motion Pictures* (New York: Macmillan, 1937), p. 13.
19. Ibid., p. 100.
20. R. Moley, *The Hays Office* (Indianapolis: Bobbs-Merrill, 1945), p. 71.
21. All details concerning the history of the Production code in this paragraph are taken from Moley's *The Hays Office*.
22. R. Maltby, '"Baby Face" or How Joe Breen Made Barbara Stanwyck Atone for Causing the Wall Street Crash', *Screen*, 27, 2 (March–April, 1986), pp. 22–45.

23. N. Roddick, *A New Deal in Entertainment* (London: British Film Institute, 1983), p. 39.
24. D. Bordwell, J. Staiger and K. Thompson, *The Classical Hollywood Cinema* (London: Routledge and Kegan Paul, 1985), p. 343.
25. Ibid.
26. Except where noted all details in the following paragraph are taken form Barry Salt's *Film Style and Technology* (London: Starword, 1983), pp. 256–7.
27. P. Ogle, 'Technological and Aesthetic Influences on the Development of Deep Focus Cinematography in the United States', in B. Nichols (ed.), *Movies and Methods*, Vol. 2 (London: University of California Press, 1985), p. 65.
28. D. Bordwell, *et al.*, op. cit., p. 343.
29. P. Ogle, op. cit., p. 66.
30. Ibid.
31. F. E. Basten, *Glorious Technicolor* (London: Thomas Yoseloff, 1980), p. 47.
32. Ibid., pp. 170–1.
33. R. Fumento (ed.), *42nd Street* (London: University of Wisconsin Press, 1980), pp. 21 and 39.
34. Details of awards taken from C. Steinberg's *Reel Facts* (Harmondsworth: Penguin, 1981).
35. J. Baxter, *Hollywood in the Thirties* (London: Tantivy Press, 1968), p. 26.
36. T. Rainsberger, *James Wong Howe, Cinematographer* (London: Tantivy Press, 1981), p. 174.
37. Ibid., p. 175.
38. Details of awards taken from Steinberg's *Reel Facts.*
39. T. Rainsberger, op. cit., p. 66.
40. Details of awards taken from Steinberg's *Reel Facts.*

7 SCREWBALL RESTRAINT: *THE AWFUL TRUTH*

1. Details on Columbia's history are taken from Chapter 8 of Douglas Gomery's *The Hollywood Studio System* (London: Macmillan, 1986).
2. L. A. Poague, *The Hollywood Professionals, Vol. 7: Wilder and McCarey* (London: Tantivy Press, 1980), p. 224.
3. Ibid., p. 308.
4. G. Weales, *Canned Goods as Caviar* (London: University of Chicago Press, 1985), p. 140.
5. D. Gomery, op. cit., p. 167.
6. Details of awards and box office success from C. Steinberg's *Reel Facts* (Harmondsworth: Penguin, 1981).
7. M. Thorp, *America at the Movies* (New York: Arno Press, 1970), p. 8.
8. Ibid., p. 10.
9. A. Bergman, *We're in the Money* (London: Harper and Row, 1972), p. 132.
10. K. Reader, *The Cinema: a History* (London: Hodder and Stoughton, 1979), p. 84.
11. Ibid., p. 71.
12. S. Cavell, *Pursuits of Happiness* (London: Harvard University Press, 1981), p. 1.
13. A. Bergman, op. cit., p. 133.
14. D. Byrge and R. M. Miller, *The Screwball Comedy Films* (London: St James Press, 1991), p. 4.
15. W. Gehring, *Screwball Comedy* (London: Greenwood Press, 1986), pp. 4–6.
16. Ibid., p. 70.

17. Ibid., p. 52.
18. A. Britton, *Cary Grant: Comedy and Male Desire* (Newcastle: Tyneside Cinema, 1983), p. 9.
19. L. Jacobs, *The Rise of the American Film* (New York: Teachers College Press, 1968), p. 535.

8 THE RESTRAINED STYLE, 1936–38

1. Details of studio history in the following section are taken from Douglas Gomery's *The Hollywood Studio System* (London: Macmillan, 1986) and from Tino Balio's *United Artists* (London: University of Wisconsin Press, 1976).
2. J. Baxter, *The Cinema of Josef von Sternberg* (London: Zwemmer, 1971), p. 128.
3. J. Baxter, *Hollywood in the Thirties* (London: Tantivy Press, 1968), p. 26.
4. D. Gomery, op. cit., p. 70.
5. Ibid., p. 86.
6. Except where otherwise noted, production details are taken from R. Behlmer (ed.), *The Adventures of Robin Hood* (London: University of Wisconsin Press, 1979).
7. F. E. Basten, *Glorious Technicolor* (London: Thomas Yoseloff, 1980), p. 66.
8. Details of box-office success and awards from C. Steinberg's *Reel Facts* (Harmondsworth: Penguin, 1981).
9. Details of awards as above.
10. F. S. Fitzgerald, *Three Comrades* (New York: Popular Library, 1978), p. 232.
11. M. A. Doane, *The Desire to Desire* (London: Macmillan, 1988), p. 63.
12. I. R. Hark, 'The Visual Politics of The Adventures of Robin Hood', *Journal of Popular Film*, 5, 5 (1976), p. 3.
13. Ibid., p. 6.
14. A. Bergman, *We're in the Money* (London: Harper and Row, 1972), p. 169.
15. N. Roddick, *A New Deal in Entertainment* (London: British Film Institute, 1983), p. 122.
16. G. Jowett, *Film: the Democratic Art* (Boston: Little, Brown and Company, 1976), pp. 282–3.

9 TOWARDS *FILM NOIR*: *DEAD END*

1. Unless otherwise noted, all information on United Artists is taken from T. Balio's *United Artists* (London: University of Wisconsin Press, 1976).
2. As William Wyler notes in A. Madsen, *William Wyler* (London: W. H. Allen, 1974), p. 126.
3. A. Madsen, op. cit., p. 155.
4. B. F. Dick, *Hellman in Hollywood* (London: Associated University Presses, 1982), p. 50.
5. Ibid., p. 52.
6. A. Madsen, op. cit., pp. 155–6.
7. Ibid., p. 156.
8. M. A. Anderegg, *William Wyler* (Boston: Twayne, 1979), p. 63.
9. A. Madsen, op. cit., p. 156.
10. Details of awards and box-office success taken from C. Steinberg's *Reel Facts* (Harmondsworth: Penguin, 1981).

11. A. Madsen, op. cit., p. 157.
12. B. Salt, *Film Style and Technology* (London: Starword, 1983), p. 282.
13. A. Madsen, op. cit., p. 137.

10 THE END OF THE DECADE

1. Statistics taken from US Bureau of the Census, *Historical Statistics of the United States* (Washington: Bureau of the Census, 1975), pp. 126 (unemployment), 224 (GNP), 226 (economic growth) and 400 (film statistics).
2. Except when otherwise noted, historical details are taken from M. A. Jones, *The Limits of Liberty* (Oxford: Oxford University Press, 1983).
3. W. Miller, *A New History of the United States* (London: Paladin, 1970), p. 344.
4. M. A. Jones, op. cit., p. 491.
5. P. Roffman and J. Purdy, *The Hollywood Social Problem Film* (Bloomington: Indiana University Press, 1981), p. 212.
6. N. Roddick, *A New Deal in Entertainment* (London: British Film Institute, 1983), p. 67.
7. C. Clarens, *Horror Movies* (London: Secker and Warburg, 1968), p. 124.
8. A. Bazin, *What is Cinema*? Vol. 1 (London: University of California Press, 1967), p. 36.
9. P. Schrader, 'Notes on Film Noir', in B. K. Grant (ed.), *Film Genre Reader* (Austin: University of Texas Press, 1986), p. 171.
10. M. B. Haralovich, 'Sherlock Holmes: Genre and Industrial Practice', *Journal of the University Film Association*, 31, 2 (Spring 1979), pp. 53 and 57.
11. R. Flamini, *Scarlett, Rhett, and a Cast of Thousands* (London: Andre Deutsch, 1976). All production details in the following paragraph are taken from this source.
12. Details of awards are taken from C. Steinberg's *Reel Facts* (Harmondsworth: Penguin, 1981).
13. R. Flamini, op. cit., p. 331.
14. Ibid., pp. 314 and 333.
15. Ibid., p. 272.

CONCLUSION

1. B. Lucich, 'The Lux Radio Theatre' in L. W. Lichty and M. C. Topping (eds), *American Broadcasting* (New York: Hastings House, 1975), p. 391.

Select Bibliography

PART 1: IDEOLOGY AND FILM

Abercrombie, N., Hill, S. and Turner, B. S. (eds), *Dominant Ideologies* (London: Unwin Hyman, 1990).
Barthes, R., *S/Z* (New York: Hill and Wang, 1974).
Belsey, C., *Critical Practice* (London: Methuen, 1980).
Biskind, P., *Seeing Is Believing* (London: Pluto Press, 1984).
Cormack, M., *Ideology* (London: Batsford, 1992).
Eagleton, T., *Ideology: An Introduction* (London: Verso, 1991).
Harvey, S., *May '68 and Film Culture* (London: British Film Institute, 1978).
Heath, S., *Questions of Cinema* (London: Macmillan, 1981).
Larrain, J., *The Concept of Ideology* (London: Hutchinson, 1979).
Lovell, T., *Pictures of Reality* (London: British Film Institute, 1983).
Metz, C., *Language and Cinema* (The Hague: Mouton, 1974).
Nichols, B., *Ideology and the Image* (Bloomington: Indiana University Press, 1981).
Nichols, B. (ed.), *Movies and Methods* (London: University of California Press, 1976).
Nichols, B. (ed.), *Movies and Methods*, Vol. 2 (London: University of California Press, 1985).
Rosen, P. (ed.), *Narrative, Apparatus, Ideology* (New York: Columbia University Press, 1986).
Silverman, K., *The Subject of Semiotics* (New York: Oxford University Press, 1983).
Thompson, J. B., *Ideology and Modern Culture* (London: Polity Press, 1990).
Wolff, J., *The Social Production of Art* (London: Macmillan, 1981).

PART 2: HOLLYWOOD IN THE THIRTIES

Balio, T. (ed.), *The American Film Industry*, rev. edn, (London: University of Wisconsin Press, 1985).
Baxter, J., *Hollywood in the Thirties* (London: Tantivy, 1968).
Bergman, A., *We're in the Money* (London: Harper and Row, 1972).
Bordwell, D., Staiger, J. and Thompson, K., *The Classical Hollywood Cinema* (London: Routledge and Kegan Paul, 1985).
Dale, E., *The Content of Motion Pictures* (New York: Macmillan, 1935).
Dooley, R., *From Scarface to Scarlett* (London: Harcourt Brace Jovanovich, 1981).
Dyer, R., *Stars* (London: British Film Institute, 1979).
Forman, H. J., *Our Movie Made Children* (New York: Macmillan, 1933).
Gomery, D., *The Hollywood Studio System* (London: Macmillan, 1986).
Higham, C., *Hollywood Cameramen* (London: Thames and Hudson, 1970).
Izod, J., *Hollywood and the Box Office, 1895–1986* (London: Macmillan, 1988).
Jacobs, L., *The Rise of the American Film* (New York: Teachers College Press, 1968).
Jowett, G., *Film: the Democratic Art* (Boston: Little, Brown and Company, 1976).
Kerr, P. (ed.), *The Hollywood Film Industry* (London: Routledge and Kegan Paul, 1986).

Kindem, G. (ed.), *The American Movie Industry* (Carbondale: Southern Illinois University Press, 1982).
Maltin, L., *The Art of the Cinematographer* (New York: Dover, 1978).
Moley, R. *Are We Movie Made?* (New York: Macy-Masius, 1938).
Moley, R., *The Hays Office* (Indianapolis: Bobbs-Merrill, 1945).
Paine, J., *The Simplification of American Life* (New York: Arno Press, 1977).
Quigley, M., *Decency in Motion Pictures* (New York: Macmillan, 1937).
Roddick, N., *A New Deal in Entertainment* (London: British Film Institute, 1983).
Ross, M., *Stars and Strikes* (New York: AMS Press, 1967).
Rosten, L., *Hollywood: The Movie Colony, The Movie Makers* (New York: Arno Press, 1970).
Salt, B., *Film Style and Technology* (London: Starwood, 1983).
Thorp, M., *America at the Movies* (New York: Arno Press, 1970).
Wasko, J., *Movies and Money* (Norwood: Ablex, 1982).

Index